GERMANY

FRANCE

ENGLAND

JURA, NEUCHÂTEL AND UNITED STATES

JAPAN AND THE FAR EAST

Japan

Beginning of the Machine Age

Height of Craft Manufacture

Geneva

Jura

USA

1600

1700

1800

1900

2000

Watch

Geneva
y Guilds

f Nantes (1598)

Years' War (1618–48)

Balance Cock (1620)

Pendulum (1657)

Fusée Chain (1670)

Wheel-Cutting Engine (1672)

Balance Spring (1675)

evocation Edict of Nantes (1685)

Waistcoat

First Jura Watch

Cylinder Escapement

Bored-Ruby Jewel (1715)

Lépine

Rack Lever

Japy (1771)

Breguet

Lever Escapement

Fontemelon (1793)

Industrial Revolution (Europe)

Railroads (1820s)

Vacheron & Constantin

Japy & Fontemelon (1845)

Stem Winding introduced

Lever accepted for quality

Pin Set (1870)

Stem Winding accepted

US Mass Production

Decline of French Ébauches

World War I

World War II

True interchangeability (1930)

Beginning of Quartz Age

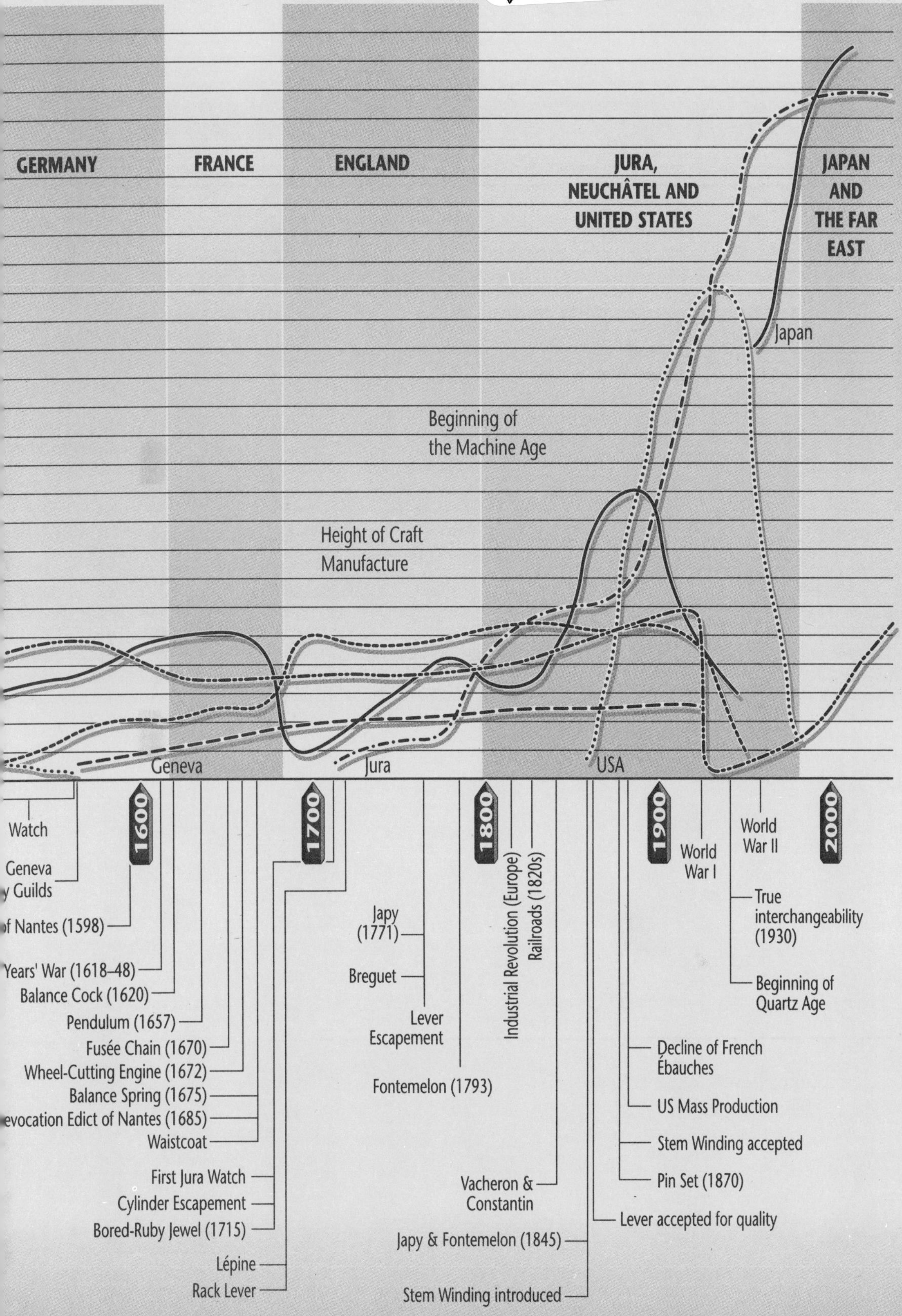

TIMEPIECES

MASTERPIECES OF CHRONOMETRY

TIMEPIECES

MASTERPIECES OF CHRONOMETRY

David Christianson

FIREFLY BOOKS

Published by Firefly Books Limited, 2002

First Printing
National Library of Canada
Cataloguing in Publication Data
Christianson, David
Timepieces : masterpieces of chronometry / David Christianson.
Includes index.
ISBN 1-55297-654-8
1. Clocks and watches—History. I. Title.
TS542.C57 2002 681.1'12 C2002-901650-9

Publisher Cataloging-in-Publication Data (U.S.)
Christianson, David.
Timepieces : masters of chronometry / David Christianson.—1st ed.
[160] p. : ill. (some col.) ; cm.
Includes index.
Summary: History of timepieces, clocks and watches.
ISBN 1-55297-654-8
1. Chronometers. 2. Clocks and watches. I. Title.
681.118 21 CIP TS542.C555 2002

Published in Canada in 2002 by
Firefly Books Ltd
3680 Victoria Park Avenue
Toronto, Ontario
M2H 3K1

Published in the United States in 2002 by
Firefly Books (U.S.) Inc.
P.O. Box 1338, Ellicott Station
Buffalo, New York 14205

This book was designed and produced by
Quintet Publishing Limited
6 Blundell Street
London N7 7BH

Senior Project Editor: Debbie Foy
Copy Editor: Andrew Armitage
Design: Ian Hunt and Chloe Garrow
Illustration: Richard Burgess
Managing Editor: Diana Steedman
Creative Director: Richard Dewing
Publisher: Oliver Salzmann

Manufactured in Singapore by Universal Graphics
Printed in Singapore by Star Standard Industries (PTE) Ltd.

CONTENTS

INTRODUCTION

THE CELESTIAL CLOCK

"Time, whose tooth gnaws away
at everything else,
is powerless against truth."

THOMAS HUXLEY

ABOVE
Early humans did not need an earthly timekeeper. The natural rhythms of the seasons told them when to plant and when to gather in the harvest for the coming winter. To plan their annual activities they relied upon the celestial clock, the natural timekeeper in the sky.

The natural rhythms of the earth—the day as it turned into night and then turned to daytime again, the sun as it tracked across the sky, even the moon as it reappeared on a regular and predictable basis—showed to early man that time was a continuous, even flow of predictable events. Even the seasons of humanity—birth, maturity, old age and death—were constant and predictable.

But at one time the concept of time meant nothing to humans. Their days revolved around the natural rhythms of daylight and dark; springtime and fall. Nature told them when to sleep and when to work; when to plant and when to harvest; when to hunt and when to gather fuel and food for that long cold season that came after the season of falling leaves. Yet even in such a simple society, the hunter-gatherer found a need to pinpoint specific moments on this continuous, flowing line of time.

If they wished to meet another to help with a hunt, for example, they needed to identify a particular moment on the line of time, so they looked for ways to isolate and communicate this reference point. They noted that the shadow of a tree grew progressively shorter as the sun rose to midday, and then grew longer on the other side of the tree as evening approached. They knew from previous experience that, as the sun peaked lower and lower in the sky, the cold season was approaching, and, after so many waxings and wanings of the moon, spring would again appear.

So, with the need to interact with others and to prepare for the long winter ahead, humans learned early on to use and rely upon the sun and moon to map out their days and plan for the foreseeable future. They learned to use the *celestial clock*.

A CLOCK IN THE HEAVENS

The celestial clock is that natural timekeeper in the sky that tells us the passage of time: the days, the months and the seasons of the year.

ABOVE
In the celestial clock the sun is the hour hand, casting a shadow as it passes from east to west. This is an example of a primitive shadow clock from Qus, Egypt, similar to those of ancient Egypt in the tenth century B.C. *The board is placed in an east-west direction and the passage of time is measured in intervals by the shadow cast on the horizontal piece of wood.*

The ancients imagined a dome above the earth upon which the sun, the moon, the planets and the stars moved. Later, when we learned that the earth was round, the dome became a sphere.

The hour is an interval of time. Its length and the number of hours in a day varied from culture to culture. But generally the daytime was divided into hours and the nighttime into periods of watches, when the guard would change in the towns and villages. Ancient Egyptians were probably the first to separate the day and night into 24-hour periods. This concept of the hour slowly spread throughout the Greek and Roman empires. The hours were called "unequal" or "seasonal" hours because their lengths varied with the changing seasons and between night and day, their length being tied directly to the length of the *solar day*.

In the celestial clock, the sun is the hour hand. As it passes through the heavens from east to west, it casts a shadow that moves from west to east. When a stick was inserted at an angle into the ground, it became the first clock made by man to tell the time: the shadow clock cast a shadow that moved in a semicircle around the stick and the shadow was divided into even intervals of time by markers placed within its path.

From this evolved the simple sundial with its angled style (the *gnomon*) casting a shadow upon a base that had the hours marked upon its face. Being made first from carved stone and then from iron and later brass, sundials were not only small individual timekeepers, but were installed upon the faces of buildings to act as public timekeepers, too.

RIGHT
To measure events, early man devised portable time-sticks such as this Tibetan priest's time-stick from Darjeeling.

The passage of the sun also tells the time of year. As the seasons pass, the path the sun takes across the sky moves each day, drifting to the north in the spring and summer months and to the south in the fall and winter (for those of us living in the Northern Hemisphere). In the summer, its path is higher in the sky. In the winter, the sun's path is lower, giving shorter days as it slips below the horizon more quickly than it does in summer.

The sundial evolved to take advantage of this seasonal change in the sun's path, to show the time of day by the angle of the shadow and then the approximate date by the shadow's length.

The monthly cycle of the moon, along with its journey through the 12 constellations of the zodiac, marks out the months and becomes the calendar function of the celestial clock. Sundials indicate the tides and tell of the seasons, predicting the times for successful sailing, planting and harvesting—all patterns our ancestors needed to know. They also allowed humans to track events and anticipate events of longer duration than a day or a week.

ABOVE
Known as the London Byzantine Sundial-Calendar, these fragments are all that remain of a portable sundial with a geared calendar. Dated around A.D. 250, these gear wheels are the second-oldest examples of gearing known and represent the early efforts of the ancients to mechanize, measure and anticipate the movements of the sun and moon across the heavens.

EARTHLY TIME

Overcast skies and long winter nights enticed man to measure time on earth by emulating the heavenly cycles of the celestial clock. Since time is a continuous, even flow of events, it stands to reason that man should be able to measure time if he has some device that will move at a continuous, even rate. A damp, smoldering rope, for example, would burn at a relatively even rate. With a knot at each hour's interval, the burning rope would indicate the time since the start of the burn. A burning candle marked with bands would likewise indicate an hour's duration between bands or a burning oil lamp could be made to do the same, with its oil reservoir marked to indicate the hours. We are all familiar with the hourglass, filled with sand that trickles with a regular, consistent flow from one side of the glass to the other. Then there was the idea of a pail of water pierced with a hole through which

a stream of water flowed. This, too, would be marked with indications to show the time as the level of water fell.

Each had its limitations. The burning rope had to be awfully long to measure an event of any length, and it needed constant attention to make sure that it continued to burn. The oil lamp, too, needed almost constant attention. The hourglass could be made large enough to measure longer events, but, at these larger sizes, the sand tended to clump and clog the orifice through which it was supposed to flow so evenly. The water clock would freeze in winter, but it was quite reliable in the milder months and in the regions of the world that were free of freezing weather.

ABOVE
Since time is a continuous and even flow of events, a vessel of water, pierced with a hole through which a stream of water flows, could be used to measure the passing of time. This example of a water clock, or clepsydra, *was used to measure time in the Athenian courts of justice in the fifth century,* B.C.

Plato (*c.*428–*c.*347 B.C.), the Greek educator and philosopher, is credited with inventing the water clock, and it had the very practical application of regulating the length of speeches in the law courts of Athens! Improvements in the design had a floating figure point to the hour, or had the dripping water turn a small wheel, which connected it with a hand on a dial to show the time. A water clock was known as a *clepsydra* and could be made in any size, from a small one for personal use to a large tower clock using buckets and water wheels for public viewing.

Muslims were concerned to know the time for their religious practices, but they loved to build water clocks that told the hour by sound and animation. Long before medieval Europe could do so, Muslim clockmakers added bells, gongs and moving figures, all choreographed by intricate gearing and regulated by the flowing water from their clepsydra. In fact, when the caliph of Baghdad wanted to impress Emperor Charlemagne he sent one of his magnificent clepsydras, as a gift and to show the Franks his nation's technological superiority at a time when the Frankish empire controlled all of Western Europe and was impinging upon his Islamic borders.

SUN TIME AND CLOCK TIME

With the advent of uniformly running timekeepers, such as burning candles and clepsydras, we soon discovered that there was a difference between *sun time* (or solar time) shown by the sundial and *clock time.*

Solar time marked uneven hours. Depending upon the time of year, the 12 daylight hours may be longer that the 12 nighttime hours and vice versa. These daylight hours gradually lengthened and shortened throughout the year (unless you lived at the equator). The continuous, even flow of the clepsydra marked "even" hours. The 24 hours of the day were equal, regardless of whether they were in daylight or darkness, summer or winter.

All of us, without exception, live by nature's clock. Daytime, nighttime, each season of the year—they all repeat themselves in a rhythm that is imprinted on nearly everything we do. These rhythms correspond with our societal endeavors—day for work, night for sleep; the sequence of seasons for planting, growth, harvest and rest—only to be repeated again and again.

To the majority of people living in a rural atmosphere, the idea of uniform (or equal) time meant nothing as far as work and sleep, planting and harvesting went. But to city dwellers, away from the animals that would awaken them at dawn and away from the need to plant and nurture the ground, the idea of measured, equal time had its appeal. Instead of living life on nature's timeline, city dwellers arranged their lives on an abstract line, with points designated as hours and minutes. Jobs would begin at a specified point on the timeline, whether it was light or dark outside. Appointments would be fixed on

BELOW
The ancient Egyptians were probably the first to divide daytime and nighttime into 12 similar spans, each of which became known as an hour. Because these hours vary in length from season to season, devices such as this horary quadrant, dated 1399 A.D., were developed to show the varying lengths of hours for each month and day of the year, as measured by the angle of the sun above the horizon.

this timeline, as would buying, selling, transporting, and distributing goods and services—an artificial time based upon hours and minutes, regulated by the even, continuous flow of manmade time.

ASTRONOMER-PRIESTS

Ancient priests were the first to study the stars and the periodic phenomena that appeared in the heavens—eclipses, meteor showers, shooting stars and comets, for example—because they thought that the heavenly bodies had an influence on the course of human affairs. With patient observation and a reliance on measured (or uniform) time, these early astronomers discovered that such mysterious events occurred at predictable times.

ABOVE
Every season has its constellations of stars. The sun moves through the visible band of stars made up of the 12 constellations of the Zodiac. Telescopes helped astronomers plot the season, the day and even the hour from observations of the stars in relation to the earth. This painting depicts sixteenth century astronomers from India viewing the stars through a telescope while consulting an astrological chart of the Zodiac.

Events that mystified and often frightened casual observers—such as a total eclipse of the sun in broad daylight or a brilliant comet apparently headed directly for earth—could be predicted in advance by astronomer-priests whose powers to (seemingly) control celestial mysteries had the desired effect on the ignorant populace. The astronomer-priests sought to create a calendar so that these heavenly events, along with the numerous journeys of planets and stars through the zodiac, could be calculated on the basis of a solar year. Not only did they know when certain astrological and astronomical events would occur but they could also find their own position in time within the solar year.

The waxing and waning of the moon, from new moon through full moon to the beginning of the next new moon, takes a month—almost. Actually, the moon's journey around the earth takes, on average, $29^1/_2$ days. It being almost 30 days, one would think that a year would have about 360 days in it. But the year is not the 12 revolutions of the moon around the earth. A year is one revolution of the earth around the sun, from the start of one season to the start of the next same season.

OBSERVING TIME

The day is our first measured unit of time, but there are various kinds of day, all based on one rotation of the earth with reference to some point *outside* the earth. This point may be the sun, an imaginary sun, a particular star or even an imaginary point in the sky.

As an observer turns with the earth, he thinks of himself as an index mark, much like the mark on a sundial, and as the sun makes two successive passes over him at high noon, the time between the two passes is an apparent solar day. Because the path of the earth around the sun is elliptical, and because the ellipse is tilted relative to the equator, these solar days vary throughout the year. If we were to put the earth in a circular path around an imaginary sun, then the solar days would all be equal to each other and would show *uniform mean solar time*, just as a good clock shows.

ABOVE
The constant, even ticking of a clock was used to track the passage of time. Shown here is a Nuremberg table clock, dated circa 1530. It features not only a hand to indicate the hour, but also a hand to show solar time and lunar time, and a disk to show the current phase of the moon.

Mean solar time (clock time) and *apparent* solar time (sundial time), however, vary with the season of the year. In the middle of April, the middle of June, the end of August, and around Christmas time, mean solar time and sundial time are equal. At the beginning of November a sundial is about 16 minutes ahead of clock time, but in the middle of February it is 14 minutes slower. The differences for every day in the year are tabulated in almanacs as the "equation of time."

The variations in solar and lunar time as compared to equal or average solar time explain why we began using star time (*sidereal* time) as a standard time reference. If an observer on earth notes the passage of a particular fixed star overhead at the same point on two successive nights, he will note that this passage is practically the same no matter what season of the year it is. A sidereal day is shorter than an average solar day (or clock day) by 3 minutes and 56 seconds. This adds up to 24 hours slower in a year. So there are 366 sidereal days compared with 365 ordinary solar days in a year. It was this sidereal day (23 hours, 56 minutes and 4.1 seconds long) that was considered the absolute time standard for centuries until the twentieth century, when mechanical clocks became so accurate that they could measure the wobble in the earth's rotation and the drifting of stars. Until then we had thought their movements were stable and unvarying.

ASTRONOMICAL TIME

Astronomers were far more interested in mean solar time and sidereal time than were ordinary people. The study of the heavens itself required an accurate time standard. The significance of all of this for ordinary people does not become apparent until they get to the point of living in communities and needing to use time to regulate every activity, not only in hours and minutes but to locate themselves within the year as well—a point that will become all too apparent as we follow the story of timekeeping.

To the ancients—and not-so-ancients—the heavens were filled with mystery, superstition, mythology and magic when certain planets aligned with certain stars or constellations of stars. All had names and each was deemed to have power over humans on earth. It was the astronomer's job to follow these celestial events, to anticipate them and predict their eventual outcome, so that the kings, priests, emperors and ordinary citizens could appease their will and act accordingly. To this end the astronomer developed charts of the earth and the luminaries that surrounded it: the sun, the moon and the visible planets and stars.

Astronomers developed hand-driven, three-dimensional models of their celestial charts, called *armillary spheres*, to add motion to their charts. With these they could study the relationships involved. The armillary sphere was an elegant machine with bands tracing out the paths of visible planets and stars, along with moving rings of horizons, ecliptics and imaginary reference points. With the earth inside the cage of rings, the cage rotated in the sphere of immovable bands, imitating

ABOVE
The astrolabe is the most sophisticated of the early time-telling instruments. This all-in-one tool shows a 12-dimensional representation of the heavens in their daily rotation. Positioning the altitude of the sun or a selected star across the star chart, one can calculate the time of day, as well as the day and month. This early example is an English astrolabe, dated 1342.

the sky turning above us and showing how the heavens rise in the east and sink below the horizon in the west.

Surrounding it all are the constellations seen throughout the year, appropriately inclined to the horizon ring so that an observer can understand the rising and setting points of the sun and moon during their paths and so understand how the seasons change.

It was to the armillary sphere that astronomers sought to add a uniform power supply to operate this model of the celestial chart and give a continuously running model that would duplicate the movement of the heavens in real time, helping the communities to

LEFT
This armillary sphere from a seventeenth century atlas, by the Dutchman, Joan Blaur (1596–1673) is an astronomical model of the heavens using solid rings, all circles of a solid sphere, with the earth in its center. A ring representing the earth's horizon encircles the model and a wide band—the elliptic—shows the path taken by the 12 constellations of the Zodiac, the sun and the moon around the earth.

organize their activities. The armillary sphere was also the forerunner for the development of gears and springs. (See The Mysterious *Antikythera*, page 29.)

CULTURAL TIME

Long before the Europeans, a Chinese emperor commissioned such an armillary sphere at the turn of the eleventh century. His armillary not only reproduced the movements of the sun, moon and selected stars but also showed the hours, the quarter hours and night watches (periods of time during the night that the town guards watched over the security of the town) all powered by water and controlled by a large tower clepsydra. This was a true astronomical clock and an engineering marvel, not only in its day but predating any European astronomical clock by 400 years—and nobody outside of China knew of it!

ABOVE
Muslim clockmakers made an art of the clepsydra, or water clock, by adding bells, gongs and automata (moving figurines), all choreographed by intricate gearing controlled by the flow of the water. This illustration shows a Turkish water-powered clock with automata from the Book of Knowledge of Mechanical Procedures, *1206* B.C. *by Al-Jozari, a thirteenth-century Mesopotamian inventor.*

Given this advance, one of the great curiosities of world technological development is why the mechanical clock was developed in Europe instead of China? After all, medieval China was technologically ahead of Europe in many areas: gunpowder, accurate water clocks, paper, movable type, porcelain, the magnetic compass and a true astronomical clock. But the mechanical clock—a logical successor to the highly mechanized and complicated Chinese astronomical water clock—did not materialize. Instead, it emerged 250 years later from the virtual scientific wasteland that Europe was during the thirteenth century.

The answer, we will see, is in the cultures of the two societies. The ordinary Chinese did not need to know the time in order to know what had to be done. Natural daily rhythms were good enough. Only the imperial astronomers needed to know the time in order to study and predict the movements of the heavens. The life of the emperor and the rule of his government were regulated by the religious mythology of the stars. The clock was an adjunct to the armillary sphere, supplying power and control to the display of heavenly motion so that the government could govern and imperial fortunes and futures could be told. They did not need to know the time, only the motions of the heavens.

In Europe, Christianity was the culture. As Professor David Landes noted in his unique work, *Revolution in Time,* "The Clock did not create an interest in time measurement; the interest in time measurement led to the invention of the (mechanical) Clock." And it was the Christian Church that had the real interest in time measurement and the ordering of people's lives.

SU SUNG'S ASTRONOMICAL CLOCK (A.D. 1096)

Su Sung was a government administrator chosen by the Emperor of China to build a magnificent clock. Cast in bronze after eight years of study, research and model-making, the astronomical clock occupied a tower roughly 40 feet (12 m) tall. It was designed to reproduce the movements of the sun, moon, and specific stars that were crucial for calculating the Chinese calendar and for interpreting the fortunes hidden in the astrological signs of Chinese mythology.

It had two primary components: an armillary sphere with its series of rings representing the paths of the crucial heavenly bodies as they appeared to an observer on earth; and a celestial globe. The armillary sphere and the celestial globe rotated on a polar axis that was properly inclined to the horizon, and driven by a pair of transmission shafts. One of the shafts also drove a series of wheels that carried jacks, or figurines, that showed the hours, the quarters and the five night watches.

The entire mechanism was driven by a water wheel designed to turn at an intermittent but stable rate. The water wheel carried a series of pivoted buckets along its circumference. As a bucket was filled with water to a predetermined weight, it tripped a lever that released the wheel and allowed it to rotate, bringing the next bucket into position to be filled. As the wheel rotated from one bucket to the next, it turned the transmission shafts, the armillary sphere, the celestial globe and the jacks, duplicating the movement of the heavens, as well as showing the time of day and night. Filling the water-wheel buckets and controlling this whole complex astronomical clock was the clepsydra. It was the clepsydra, or water clock, that kept the time, parceling out the flow of water into measurable units of time.

Su Sung's magnificent astronomical clock served the Emperor and his people for a few years until invaders from northern China captured the capital and carried away the clock. From there it disappeared.

It was lost until the year 1172, when the original book (with diagrams and descriptions of the clock) was found in southern China. The original was lost again and then it finally resurfaced in the nineteenth century. In the mid-twentieth century several working models of Su Sung's clock were created from the details and descriptions found in one of the reprinted books. One model now resides in The National Time Museum of Chicago, and another in the National Museum of Science and Industry, London.

ABOVE *Su Sung's magnificent astronomical water-powered tower clock. Built in the eleventh century, the tower has an armillary sphere on top, a celestial globe inside and time-indicating figurines.*

CHAPTER 1

A CALL TO PRAYER

"Lost time is never found again."

BENJAMIN FRANKLIN

ABOVE
To the highly regulated and cloistered world of the Christian monastery, time was of the essence. It was within the monastery, and for the Christian Church as a whole, that the mechanical clock was developed to measure and regulate people's lives.

A faint peal of the bells from the monastery atop a distant hill reminded the faithful that it was time for prayer. Within the highly regulated medieval community of the monastery, the brothers, lay brothers and servants were busy everywhere. They lived and worked to the bells that called them to a daily regimen of prayer and ceremony, work and meditation. Time was of the essence because time belonged to God and it could not be wasted.

The monastery ran by strict rules and an unerring schedule. It was the duty of those in the monastery to pray, and to pray often, in order to save the multitude of faithful outside the monastery, whose worldly duties kept them from devoting their full time to the service of God. Their praying took the form of seven daytime services, called *canonical hours*, because they were recited at specified times, and one at night called the *vigil.* Organized around these "hours" were times for individual prayer, study and the other activities of cloistered life.

To the Christian, simultaneous group prayer was considered the most powerful form of prayer. Time discipline was taken seriously in the monastery, and nothing was so important as the punctual, collective prayer, chanted aloud. It was not acceptable to be late for prayer. To the Christian community, the whole was greater than the sum of the individual parts.

The Christian was different from adherents of the other religions that worshiped the same God. The Judaic worshipper is obligated to pray three times a day, but not at specific times: in the morning after the sun comes up; in the afternoon before sunset; and in the evening after dark. Nature gave the pious Jew the signals to pray—no timepiece or alarm was needed.

Islam calls for five daily prayers: at dawn (just before sunrise); just after noon when the sun is highest; sometime in the afternoon before sunset; just after sunset; and then after dark. Here again, nature gives the signals for prayer. In both Judaism and Islam, local religious leaders could set times for prayer, but these times were not absolute.

Offering daily prayers the requisite three or five times was absolutely necessary, but great leeway was given for when they were offered. After all, in both Islam and Judaism, prayer is a personal act and (with some exceptions) need not be carried out as part of a group.

Christianity, especially that practiced in the monasteries, was different. The early Christians had no standard liturgy, they were not yet even a church. But, since most of the originators of Christianity were Jews, they used the practices of the older faith to build their own. They accepted the morning and evening prayers "when thou liest down and when thou risest up" (Deuteronomy 6:7), and the three daily prayers called for by Daniel, when "he kneeled upon his knees three times a day" (Daniel 6:10). And they added additional times, making eight in all, in order to distinguish themselves from the Hebrew practitioners of the faith. As the Christian church developed, these times of prayer became a part of the church canon of law and divided the day into canonical hours around which all other activities were organized. The apostle Paul taught that the faithful should be in a state of continual prayer. By virtue of their calling, the monks were in a perpetual state of communion with God, no matter what they were doing: they made no distinction between work and prayer. To them work was prayer.

BELOW

A gilt-brass-cased astronomical compendium with its outer leather case, made by Humphrey Cole, London, in 1575. This compendium is a portable instrument with a sundial, a compass to orient the sundial and a rotating disk to demonstrate astronomical phenomena and aid in their calculation.

ORDERLINESS AND MEASURED TIME

To become organized and stay organized, the monasteries measured time. The ringing of bells alerted the cloistered community to the daily activities, calling them to prayer, to study, to work, to sleep—big bells and little bells, each with its own meaning. For hundreds of years the hourglass, clepsydra and the calibrated burning lamp were the instruments used to remind the bell-ringers when to signal the next event. To adjust for the changing daylight and nighttime hours, the clerics carefully measured and proportioned the time, calculating the length of each night and day for every day of the year, measuring to the nearest quarter-hour or so because the time-measuring devices were no more accurate than this.

Time mattered to the monks because being late for the all-important prayer was stealing time away from God; and sleeping through the nighttime vigil could have serious consequences. The apostle Matthew admonished the faithful that the Lord would come at an hour when they would not expect him (Matthew 24:42–44). And Luke added that "if he comes in the second watch, or in the third" blessed are the servants who are watching for the Second Coming of the Lord. Woe be, then, to the poor bell-ringer who failed to awaken his brethren for the nighttime vigil!

ABOVE
For thousands of years people have debated the heavens and their place in them. Early civilizations created mythologies in which the sun, moon and visible planets (formed into the mythical constellations of the Zodiac) played central roles. They ordered their lives by the regularity of the heavens but, from time to time, unusual events such as the solar eclipse observed here by French astronomers in the sixteenth century, required investigation and explanation.

AUTOMATING THE RINGING BELL

It was for the potentially hapless nighttime bell-ringer that the mechanical alarm developed. Akin to our baking timer, the alarm timer would run for hours, tripping an alarm at the appointed time and awakening the bell-ringer who would then ring the vigil bell for the rest of the monastery. These devices were not clocks in our sense of the term. They did not record the time or run continuously. They were timers.

The alarm timer that woke the bell-ringer evolved in the twelfth century into a mechanism that actually rang the bell, using a series of gear wheels that were driven or powered by a falling weight. The gears would turn a ratcheting lever that clamored against the bell. A clepsydra would determine when to release the bell-ringing mechanism. This clepsydra-controlled automated bell was ideal on a small scale, but to ring the bells on a large scale—the big bell in the monastery tower, for instance—required a different controller. It was not practical to bring water up into a tower to feed the clepsydra that controlled the large automated bells.

Inventive clerics and blacksmiths put their minds together and, in the thirteenth century, the true mechanical clock emerged—unique to Europe. Other countries had their sundials and clepsydras, but only Europe had the mechanical clock. Using the weight-driven gear wheels of the alarm mechanism, these creative geniuses realized that the back-and-forth motion of the bell-ringer could be used to control the speed of the gears. By placing small weights on both ends of the

THE CROWN-VERGE ESCAPEMENT

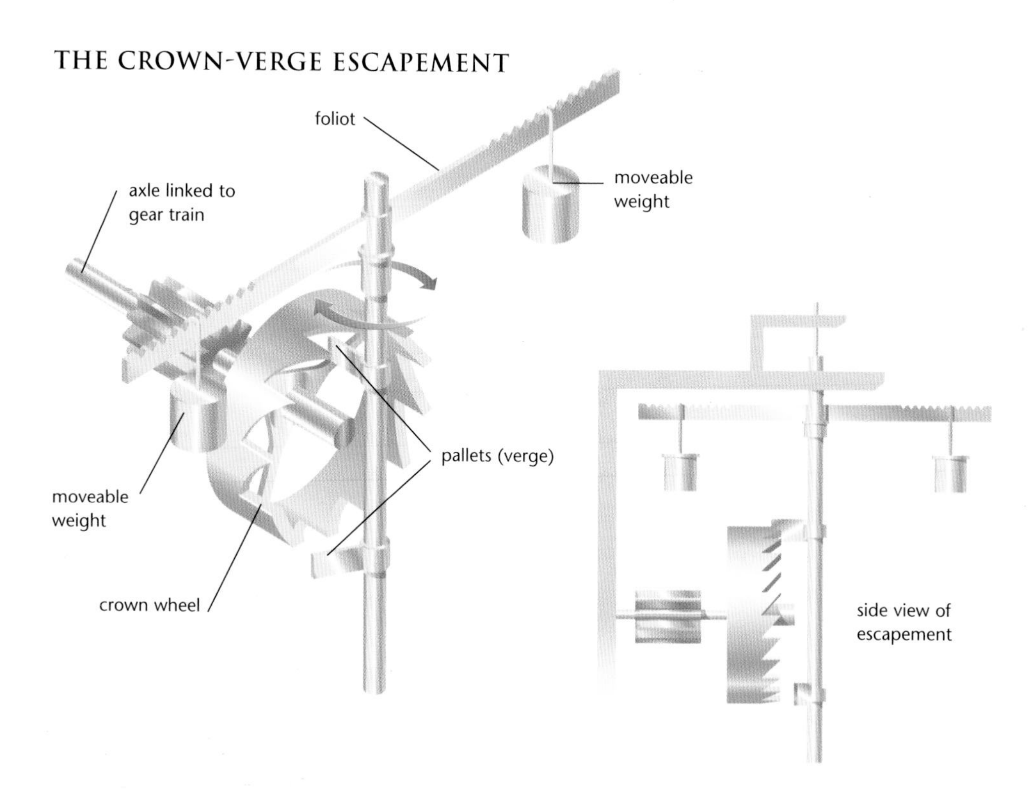

ABOVE
Two views of the crown-verge escapement with its foliot oscillator. The foliot arm oscillates back and forth, catching and releasing the revolving crown wheel one tooth at a time. As a tooth is caught by a pallet it pushes against that pallet, sending the foliot in a reverse direction, thus maintaining the foliot's oscillation. The weights on the foliot adjust the speed of oscillation and therefore regulate the speed at which the crown wheel and the adjoining gear train wheels revolve.

ringing lever, they made a speed governor that released the power of the weight-driven gear wheels at a relatively constant rate. The last wheel in the gear train is the *crown wheel* (so named because it looks like a crown). Its pointed tips, which were blocked and released by pallets on the axle of the oscillating lever, gave a back-and-forth motion to this oscillating lever arm. The pallets on the axle were called the *verge*, and the crown wheel was called the *escape wheel*, because this wheel allowed the rotating motion of the gear train to "escape" one tooth at a time. The speed of the escaping teeth was governed by the length of the lever arm, called a *foliot*, because it seemed to have a mad motion as it moved back and forth (from the Old French *fol*= mad.)

By moving the weights on the ends of the foliot, the rate of its oscillation could be changed—up or down—to bring the speed of the escaping wheels into some standard, measurable rate. The standard would be the time from sunrise to sunrise. By adding the correct ratio of gear wheels, the clockmaker could produce a series of gears, one of which would turn every 12 hours. With a lever on that wheel, it could ring a bell twice a day, and an indicator hand on the wheel would give you a clock to record and measure time. For a long time the hand was not there. The clock simply struck the bells and you knew that it was time for prayer or work or sleep by the number and sound of the bells.

CLOCKWORK GEARING

Clockworks are synonymous with the mechanical clock. Toothed wheels mesh with toothed wheels, never slipping, always turning in an inflexible sequence; each rotates on an axle or arbor, smoothly transmitting energy from a power source, such as a falling weight, through a series of gears, called a train. A gear with fewer teeth propels an adjacent wheel more slowly in circular motion than a gear of equal size with more teeth, assuming both are turning at the same speed. A smaller gear moves more quickly in circular motion than a larger one, given the same motive force. Linking the right series of wheels together, each with different rates of circular motion, one can create a clock with hands that will record the passage of time. Or one can create a mechanical calculator that can track the sun through the zodiac and trace the epicyclical orbits of the planets—all using a gear train that can be reduced to precise mathematical relationships that would become known as clockwork. Clockwork gearing had its origins in Greece in the third century B.C. The toothed gear wheel appeared about the time of Archimedes (278–212 B.C.) and invariably throughout history has been linked to this mechanical and mathematical genius.

For 1,500 years the toothed gear was used to model the movements of the heavens. There is an ancient tendency to make models that duplicate natural phenomena, which is still an important part of scientific methodology today. Successful model-making is an essential element in scientific achievement. In order to understand and prove a principle, process or effect, one must make a successful model of it. In fact, science has probably developed more easily and rapidly from the scientific simulations of theories (using models) than from the practical application of these theories.

Internal clockwork gearing from the power source onward, multiplying its evolving motion until the various hands revolve at the speed of elapsing time, whether it be hourly, lunar, solar or sidereal.

POPE SYLVESTER II (A.D. c. 945–1003)

Sylvester II was the first French Pope. It was his ambition as Pope to unite all of Western Europe into one church and one state. Before he became the "Pope of the Millennium," he was Brother Gerbert of Aurillac, a Benedictine monk. He was truly a genius in his generation. He practiced the art of dialectic: the method of finding the truth by disclosing the contradictions in an opponent's argument—an indication of a truly brilliant, logical mind. And he had studied mathematics and astronomy in Spain, probably at the feet of Jewish and Muslim scientists who were far ahead of those in Medieval Europe. His former student, a monk named Richer, tells us that his teacher had built a celestial globe and an armillary sphere to show the motion of the heavens. Because of this, Brother Gerbert presumably would have had the knowledge and skill to build a mechanical clock to drive it; in fact for centuries he was given credit for having built the first mechanical clock. But, if he had built it, he would have needed an oscillator as a time standard, and an escapement (see page 32). None is known to have existed at that time. Did the Church suppress it as having come from the infidels—the Muslims or Chinese—or as the prize of some pact with evil? A good question. After all, Pope Sylvester II did eventually acquire a reputation as a sorcerer and a heretic! In the end, it would be another 250 years before the first mechanical clock appeared on the European scene in the thirteenth century.

A spring-driven table clock with classical busts around the case. Created in Paris in 1545, the clock had an astrolabe on top to help reset the clock's hands periodically with the sun or a bright star.

WHEN IS A CLOCK NOT A CLOCK?

It is interesting to note that there is no word for "clock" as a time-measuring instrument. The word "clock" comes directly from the Middle English word "clokke," the Old French word "clocke," and the Medieval Latin word "clocca," all meaning "bell." The clock, as we know it, derived directly from the need to ring bells, both in mechanism and in name. Even the German "glocke" means bell. But a new mechanism such as the automated bell needed a new name and "clock" was it. This mechanical clock then became the controller for the automated bells. It could be made small for controlling the ringing of small bells, and large for controlling the automatic ringing of the big tower bells.

While the development of the mechanical clock can be considered akin, in terms of technical innovation and especially socioeconomic impact, to the recent development of the computer, during the Middle Ages time was the only aspect of science that moved ahead. In other areas of science and the humanities, knowledge had stood still or was even lost. Nevertheless, time measurement was a subject of great interest. Not only did the Church need to order and regulate its every activity, it needed to find and systematize the dating of Easter and the other movable feasts. These dates were tied to the lunar cycles and the solar calendar, both of which were different, and each varied from year to year. This led to calendrical science with very complex observations and computations, all tied to time measurement and all dependent upon a reliable, consistent means of measuring time.

The clergy may have been the primary market for the newly emerging mechanical clock and hastened its technical advance, but urban centers were developing as well. Sleepy villages were becoming busy centers of social administration, commerce and trade, based on the newly expanding agricultural and commercial growth of the twelfth to the fourteenth centuries. These activities required the organization and coordination of many diverse groups and individuals to receive, package, distribute and market the goods and services coming into the cities and towns. Just like the monastery, the people involved in commerce and trade depended upon a reliable source of time measurement to put their private and working lives into order.

With commerce and trade, communities accumulated resources and wealth, and with this wealth mechanical clocks appeared in the cities

ABOVE
A typical house clock, made in 1596 by the brothers Ulrich and Andreas Liechti of Winterthur, Switzerland. The Liechti family were eminent clockmakers in Winterthur from 1514 to 1741. The clockwork's frame and dial plate are made entirely of iron, which was the state of the art in the late sixteenth century.

THE DE DONDI ASTRARIUM (A.D. 1364)

Less than a century after the development of the mechanical clock, the abbot Richard of Wallingford (*c.*1292–1336) and Giovanni de Dondi (1318–89) produced two highly complicated astronomical mechanisms (*astraria*) driven by mechanical clocks.

Richard of Wallingford's clock was a massive 8½ feet (2.6 m) high and was constructed between 1327 and 1335 for St. Alban's Abbey, Hertfordshire, England, to sound the abbey bell.

De Dondi's clock was a much smaller, more delicate machine, only 4 feet (1.2 m) high and 32 inches (81 cm) in diameter. It displayed the movements of the heavens on seven planetary dials around the top of the astrarium, with clockworks beneath. De Dondi's clock was declared a marvel. This remarkable pre-Renaissance clock is a great testimony to Italian genius and skill, and demonstrates how Italy would lead Europe into the Renaissance.

A fourteenth-century depiction of Richard of Wallingford, designer of the St. Alban's Abbey clock *(see page 26)**, measuring with a pair of compasses.*

No comparably intricate timepiece would be created for another 200 years.

These two clocks are thought to be the first two fully automated, mechanical astronomical machines made. Although both have disappeared over the span of years, their makers, like Su Sung so many years before, described them in such detail that we have been able to make working copies of both of them in our own day. The National Time Museum of Chicago owns the Wallingford model and the de Dondi clock. The Smithsonian, Washington D.C also holds a model of the de Dondi clock.

By our standards, the mechanical clocks that ran these astraria were crude, imprecise and unreliable instruments, but they were nonetheless enormous achievements in their time—especially given that they needed setting just once a day.

The de Dondi astrarium showed the positions of the sun, moon and the five then-known planets as they traveled through the zodiac. To do this de Dondi used epicyclic gear wheels, elliptical gear wheels and skew gears, the first known examples of these gears.

Below the dial of the sun is a 24-hour time dial flanked by tables giving the times of sunrise and sunset throughout the year. Also on this same level is a dial of the moon and an ingenious chain mechanism carrying the movable feasts of the Church year. What makes this clock even rarer is that it was made of copper and brass at a time when all clockworks were made of wrought iron.

An account written in 1389 says, "So great is the marvel that solemn astronomers come from distant countries to visit Master Giovanni and to marvel at his work... The subtle skill of Master Giovanni enabled him to make with his own hands the said clock... and he did nothing else for 16 years."

Giovanni de Dondi was a physician and teacher of astronomy at the University of Padua. He was born in 1318 and learned much of what he knew about building clocks from his father, Jacopo, who was an astronomer and clockmaker in Padua. Jacopo was awarded the surname of "del Orologio" (of the clock) for an important clock that he designed for the tower of the Carrara Palace in Padua. The title was a sign of the exceptional service the elder de Dondi had rendered as the maker of complicated clocks, an honor that was cherished by his descendants through the years. The title also announced a new profession: the horologist, or clockmaker.

The fourteenth-century Giovanni de Dondi's astrarium, with seven planetary dials, reconstructed from plans that have survived, although the original mechanism has not.

THE MYSTERIOUS ANTIKYTHERA (c. 86 B.C.)

The *Antikythera*, a highly complex clockwork mechanism, was once thought to have come to earth with alien visitors. But we now know that it has very earthly origins. It was found off the Greek island of Antikythera in 1900 (2,000 years after it was made), and it took some 70 years to discover that this first-century-B.C. machine was our earliest physical model of the heavens.

In the form of a calendar computer, it demonstrates and calculates the position of the sun and moon in relation to the 12 zodiac months; calculates the risings and settings of various bright stars and constellations; and calculates the lunar months and phases of the moon (which constitute the lunar year). And, what is more, it does this over the 19 years that it takes for the phases of the moon to recur in the same order and on the same days as in the preceding 19-year cycle.

In this mechanism we see the use of the first differential gear (see page 102). Much later, this type of gear showed in a thirteenth-century Islamic document and then in a sixteenth-century German astronomical clock. From there it was adapted by a clockmaker to an early textile spinning machine that helped to begin the Industrial Revolution, then it was adapted to the first steam motor car and finally to the rear axle of every automobile in use today.

The history of the clockwork begins not with a simple mechanical clock telling the time, but with a complex astronomical calculator. It remained to be used in astronomical models for 1,300 years until a measuring mechanism appeared that could pace the even flowing movement of the clockworks to the annual revolution of the earth around its sun.

CHAPTER 2

THE PRICELESS POSSESSIONS OF A FEW

"I recommend you to take care of the minutes; for hours will take care of themselves."

LORD CHESTERFIELD

ABOVE
Watchmaking traces its origins to the early goldsmith shops such as this Italian shop of the sixteenth century. The intricate tools and skills of the goldsmith lent themselves to the miniaturization of the larger clockwork mechanism into pocket watch size.

Early domestic clocks of the fourteenth century were merely smaller versions of the larger tower clocks that had piqued the people's interest in timekeeping. They were still made of the same iron framework and gears, but reduced in size from 8 cubic feet (0.23 m^3) to as small as 1 cubic foot (0.03 m^3). The gearing was still crudely cut and required heavy weights to drive the verge and foliot escapement (see page 22) that metered out the time.

Only the wealthiest members of society could afford these clocks, and just as well: they were poor timekeepers and very costly to make and maintain, just like their larger brothers, the tower clocks. But they were a technological step forward to a clock that would one day be suitable for every home.

In the centuries before the development of the mechanical clock, as Europe moved into the High Middle Ages, the continent saw a resurgence in population growth, increased agriculture and exploding commerce and trade. With this came taxes and duties and rising wealth for both the royal courts and the landowning middle class.

PATRONS OF CLOCKMAKING

It was the numerous rulers of Europe who took a particular interest in these wondrously ingenious machines. They sought out the better makers in the newly emerging clockmaking craft and took them into their courts, either sponsoring them or employing them to be clockmakers to the court. Employed and kept by royalty to make clocks for the courts to keep and to give away as gifts, the clockmaker could also be loaned out to other wealthy personages to make clocks for them, too. Considering it could take a clockmaker a year to design and build a clock, the fortunate maker who was selected as maker to the court had a lifetime job and an opportunity to create elaborate pieces that others could not afford to take the time to do.

THE ESCAPEMENT

The oscillator and escapement are the heart of the clock. The more consistent or stable the oscillations are and the more efficient the escapement is, the more accurate is the clock.

The oscillator tracks the passing moments. It goes by different names: a time standard, a frequency standard, a controller, a regulator.

The escapement counts the passing moments of the oscillator, called beats or pulses, by blocking and releasing the clockwork wheel train in a rhythm of equal intervals dictated by the oscillator, thereby moving the hands to indicate the passing time. An oscillator pulsing one beat per second—as in an English long-case clock—will have the escapement release the wheel gear train once per second. A hand on the escape wheel will make one revolution each minute, showing the passing of each second on a dial divided into 60 one-second segments. The wheel train will accumulate or multiply these seconds using a series of gears. One gear wheel moves at the speed of one revolution per hour and carries a minute hand showing the passing of a full hour on a dial divided with 60 minute marks. Another slower-moving wheel carries an hour hand that moves one revolution every 12 hours—or one-twelfth of a revolution each hour—showing the passing of each hour on the dial divided into 12 hour marks.

The verge-and-foliot (see page 22) was probably the first clock escapement and oscillator and formed the basis of future clock technology. The foliot, or pivoted bar, was the oscillator and the verge-and-crown wheel comprised the escapement, with the verge blocking and releasing the crown wheel as the foliot oscillated. The verge-and-crown wheel (or the crown-verge) arrangement was not a bad escapement: it did a fairly efficient job of counting the oscillations of the foliot. But the foliot was erratic, hence its French name which means "mad motion." Its motion was controlled too much by gravity. The escapement could initiate the oscillation but the foliot was dependent on gravity to return it to its point of rest before the escapement blocked its return and re-initiated the oscillation in the opposite direction. This motion relied on gravity to return the foliot to the center—its point of rest.

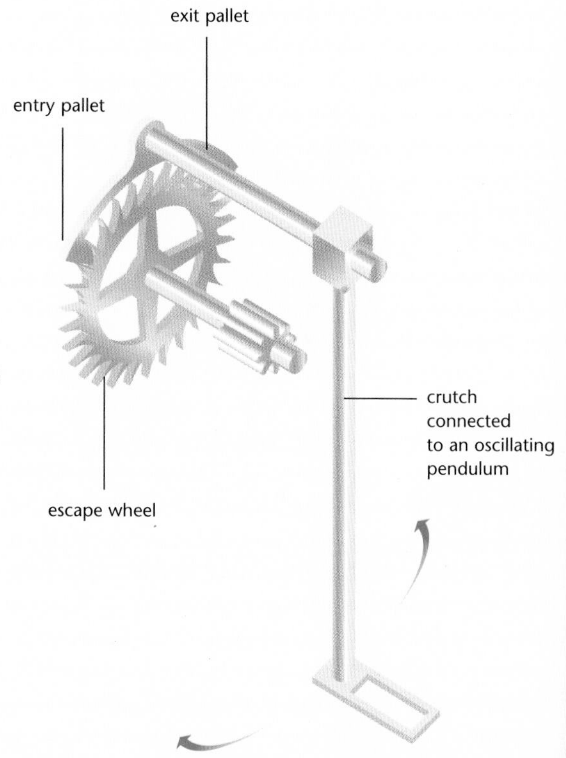

The anchor escapement, with its anchor-shaped pallets, would rock back and forth with the swing of the pendulum, releasing the escape wheel and its connecting gear train at such a speed that the second or center wheel in the gear train would turn once each hour

Sometime a little later (the actual date is unrecorded) the foliot took the shape of a horizontal wheel, the theory being that the wheel would eliminate the "mad motion" of the foliot. But this proved unsuccessful as the wheel had the same effect as the foliot. However, when a straight, flexible spring (a pig's-hair bristle, actually) was added to the horizontal wheel (called a *balance wheel*) the spring returned the wheel back from the extremes of its oscillating motion, essentially countering the force of gravity. This helped even out the "mad motion" of the oscillator. Later, when a spiral steel spring replaced the hair bristle, the *hairspring*, as it is still called, further smoothed out the erratic motion and gave the horizontal balance wheel a much more stable rate of oscillation.

The foliot was used in clocks from tower to table size. The balance wheel without a hairspring was used in some tower and wall clocks, but mainly in table clocks. With the hairspring added, it was found in carriage clocks as well as table clocks, reaching an accuracy of within a few minutes a day using the same crown-verge escapement.

The pendulum was a very stable oscillator, and when the anchor escapement was invented by Robert Hooke around 1671, (see glossary, page 162) rather than the crown-verge, the clock could be regulated to less than a minute's variation in a day. The anchor, because it operated with a much smaller arc than the verge, allowed the use of a much longer pendulum. The old verge escapement could not block and release an oscillating pendulum much longer than a few inches in size. The longer the pendulum, the more stable the oscillating rate, and from this clockmakers discovered the standard length for precision: the one-meter pendulum—called the *royal pendulum*—which beats a steady one-second rate.

The Gear Train

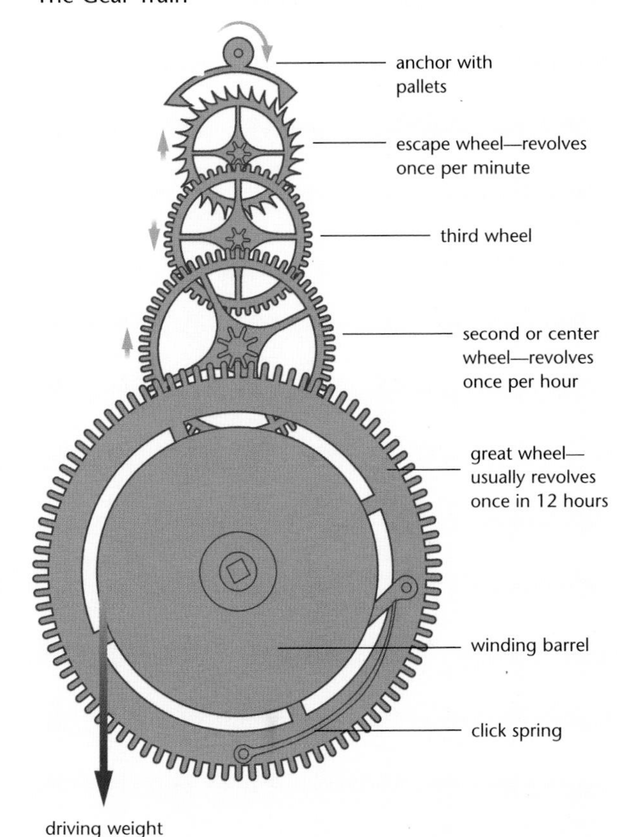

The anchor escapement releases the gear train. The great wheel drives the second wheel, enabling it to revolve one turn each hour.

RIGHT
Isaac Habrecht of Strasbourg built this miniature version of a public tower clock out of gilded brass in 1589. The clock is complete with a carillon that plays music on tuned bells every hour as moving mechanical figurines perform for the entertainment of the observer.

Royal patronage of the clockmaking art furthered the advancement of the craft. Under the employ of a royal personage, the clockmaker had time to experiment, expand his skills and further his craft. During this time brass was found to be a much finer medium than iron. Locksmiths, used to working in brass, were drawn to the craft. With brass and steel replacing wrought iron and cast iron, clocks could be further miniaturized to floor-clock size, and as clockmakers learned to make smaller and finer gears, they developed shelf clocks that could run on smaller weights with shorter falls. Complicated mechanisms, beyond timekeeping, could now be added to the clocks, such as striking bells and musical chimes, as well as animated figures that performed entertaining antics on the clock.

The flat coiled mainspring (see glossary, page 168) was soon developed. This could replace the falling weight as a power source. It required the making of even smaller and finer gear trains because the mainspring did not have as much power as a falling weight; but it did allow the clock to be moved from one room to another. The mainspring did not give an even flow of power because of the metallurgy at the time. A fully wound spring exerted more power to the gear train than a nearly relaxed one. So during this period two devices evolved to equalize the power of these springs: the *stackfreed* of Germany and the *fusée cone* (see page 36) of France.

There is a curiosity in technical development: when innovation stagnates, then ornamentation flourishes. Clocks began to take elaborate forms beyond simple wall or shelf styles. They took the form of ships, pillars, drums, animals, people and mythological characters. They were highly ornamented and elaborate in outside appearance. Owning a clock during this time was a symbol of prestige and prosperity, and giving a clock as a gift was the ultimate show of wealth and generosity. It did not matter that clocks were still such poor timekeepers; their outward appearance was far more important.

ABOVE

An animated clock in the form of a ship. The clock was created by Hans Schlottheim of Augsburg about 1585, probably for the Emperor Rudolf II. The ship rolls across the table, the sailors in the crow's nests strike the hours on bells and organ music sounds along with many other moving figures on the deck of the ship. This clock would have traveled along a banquet table for the entertainment of guests.

DEMAND FROM THE BOURGEOISIE

The wealthy landowning middle class, the bourgeoisie, gradually became a source of demand for these ingenious mechanisms and symbols of achievement. To meet the demand, the clockmaking craft grew and specialization entered the picture. While in the past a clockmaker himself built his clocks from scratch, he now employed

THE STACKFREED AND FUSÉE

The two common methods used to equalize the expanding force of the mainspring were the German *stackfreed* spring and the *fusée* used by the Italians, French and English. Referring to Figure 1, the axle that the mainspring is coiled around extends through the watch plate and has a gear on its end (A). When the mainspring is fully wound the highest part of the cam (C) is under the roller (R) of the *stackfreed* spring (S) [see Figure 1a]. This flexes the *stackfreed* spring, putting a heavy force on the cam that is affixed to wheel (B) and retarding the power of the mainspring that drives the watch. As the mainspring unwinds, the diminishing height of the cam (C) on the wheel (B) allows more of the mainspring's power to be released. As the mainspring unwinds, the roller (R) approaches the lowest point on the cam. The *stackfreed* spring (S) and roller (R) are exerting little or no pressure, and all the power in the now-weakened mainspring is being used to drive the watch [see Figure 2].

The tapered cone of the *fusée* [see Figure 3] acts like a tapered pulley connected to the mainspring barrel by a cord or chain. As the *fusée* cone is wound, the chain winds up along the cone following the spiral grooves until, at full wind, the chain is at the smallest-diameter end and the mainspring in its barrel is fully wound. As the mainspring unwinds, the chain pulls against the ever-increasing diameter of the *fusée*, equalizing the ever-weakening power of the expanding mainspring and providing nearly uniform power to run the watch.

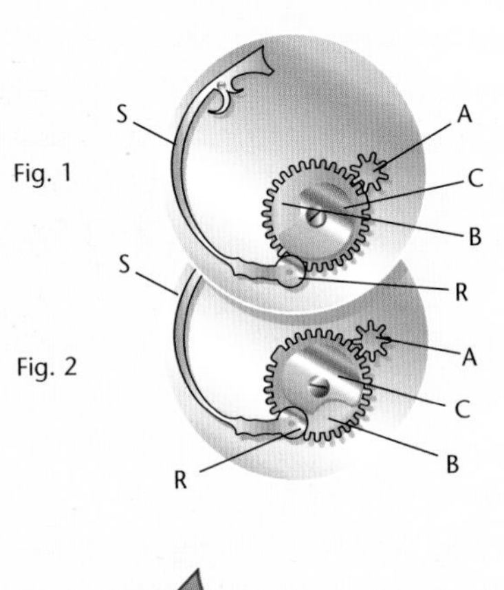

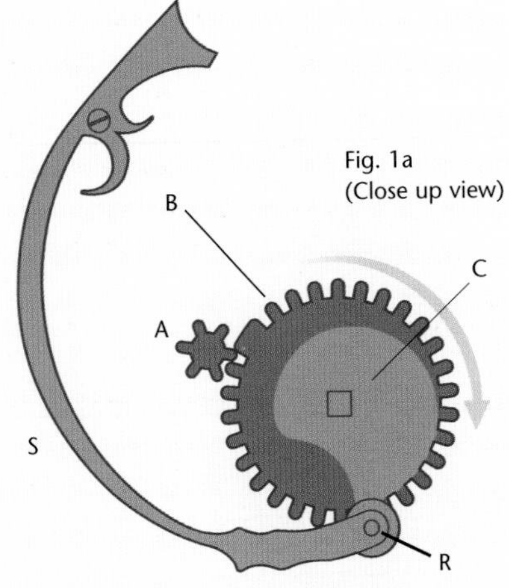

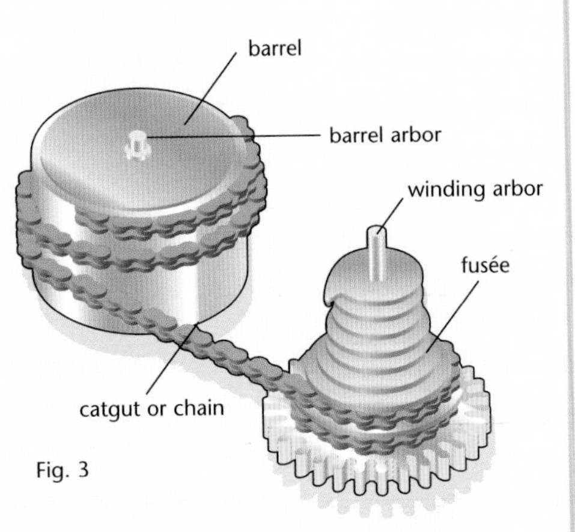

Figure 1 shows the stackfreed fully wound (close-up view, Figure 1a). Figure 2 shows the stackfreed fully unwound. Figure 3 shows the relationship between the fusée, chain and mainspring barrel.

ABOVE

A three-dimensional wood carving of a clockmaker's shop making a mechanical clock. Made by a Danish woodworker, circa 1600, the image shows the various skills represented in the shop: the machinist at his lathe in the left background; the artisan filing a part to fit (left foreground); and the designer/craftsmen in the center.

specialists in gilding and setting gemstones, and even in the building of the outer case. Specialists in making mainsprings were used, as were dial makers, hand makers and bell makers. Within the shop, employees specialized in wheel making, gear cutting, arbor making and assembling—all under the watchful eye of the master. This made sense: every worker could do some things better than he could do others, so the idea was to let each one do what he did best, all for the benefit and efficiency of the job and the shop.

A NEW DIMENSION OF ACCURACY

Some time into the mid-seventeenth century, the pendulum and balance spring (see glossary, page 162) were applied to clocks, resulting in a revolutionary influence on the future of mechanical timekeepers.

The origin of the pendulum has been traced to Leonardo da Vinci (1452–1519) who, in the late fifteenth century, made several drawings showing a gravity pendulum fitted to a clockwork mechanism. Around 1520, Benvenuto della Vopaia, a famous Italian

RIGHT

A square table clock with detachable alarm bell sitting on four legs atop the clock. The dial is on top of the clock and the sides are decorated with scenes from Orpheus. *This piece was made in Germany, probably Augsburg, around 1580.*

LEFT
This 1841 fresco depicts Galileo observing the uniform swing of a lamp suspended from the ceiling of Pisa Cathedral, Italy. From his observation, and using his pulse to time the lamp's oscillating swing, Galileo went on to develop his pendulum theory.

instrument maker, drew a clock escapement that appears to be designed to work with a pendulum.

Tradition has it that Galileo Galilei (1564–1642), while sitting in the cathedral in Pisa, noticed a great lamp swinging from the ceiling. Being the scientist that he was, he observed the swing and noticed that the lamp oscillated back and forth. Using his own pulse, he ascertained that each swing seemed to have the same duration and that the backward swing had the same duration as the forward swing. In other words, the oscillation from the center point of its swing to each extremity of its arc took the same amount of time whether it was a large arc or a short one.

Galileo returned to his studio and experimented with the pendulum, writing his observations between 1637 and 1639. Around the time of Galileo's death, his son, Vincenzio, drew a clockwork mechanism that would maintain the pendulum in motion.

SCIENCE AND THE CLOCKMAKER

The Renaissance of science began with the seventeenth century and was led particularly by Galileo. Encouraged by the creation of academies, observatories and published journals, scientists and scholars could meet throughout Europe to share their ideas and discuss their work. The need for observation and accurate measuring instruments became pressing as scientists increased their study of medicine, biology, astronomy, physics and surveying.

Learned men of the time not only made their own observations and conducted their own experiments; they made their own tools and instruments and improved upon others in order to better conduct their studies. These scientists were interested in multiple disciplines and carried out research in many diverse fields.

Leonardo da Vinci (1452–1519), the quintessential scientist long before the Renaissance of science, had an insatiable curiosity. Not only was he one of the greatest painters of the Italian Renaissance, but his achievements spread into anatomy, astronomy, botany, geology and mechanics. His notes provide us with drawings of production machines for textiles, metalwork and hydraulic machines long before they materialized. He envisioned flying machines and submersible boats, as well.

In the late fifteenth century he made several drawings of a gravity pendulum controlling a clockwork mechanism and he made a drawing of a *fusée* cone, which equalizes the driving power of a mainspring as it transmits its power to the gear train of a clock (see page 33).

Galileo Galilei of Florence (1564–1642) studied medicine, philosophy, physics and astronomy, as well as playing the lute and the organ. In addition to improving upon the telescope and discovering the law of the pendulum, Galileo invented the hydrostatic balance and designed the sector, a measuring instrument consisting of two arms marked with gradations, used by draftsmen.

Christiaan Huygens of the Netherlands (1629–95) was undoubtedly the greatest European scientist during the last half of the seventeenth century. He likewise invented or improved upon the instruments he needed for

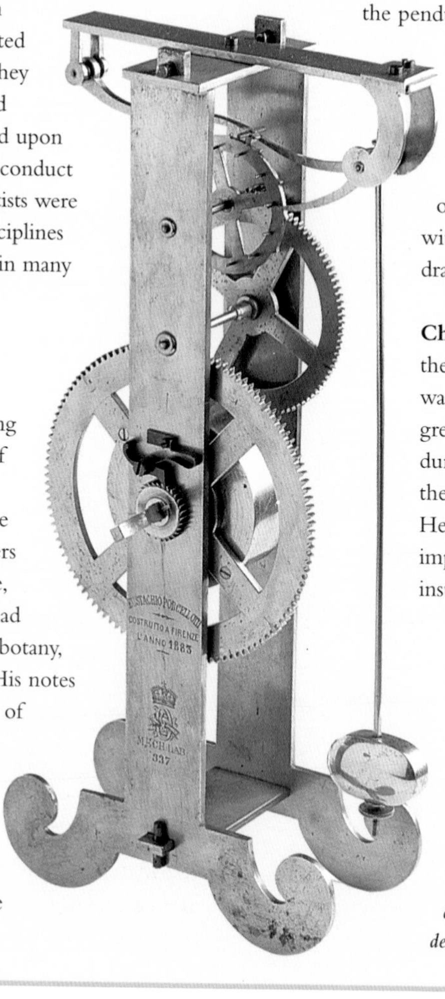

This nineteenth century model of Galileo's pendulum clock is based on a drawing of an incomplete pendulum clock that Galileo designed just before his death in 1642.

his observations and experiments, including grinding his own lenses to view the planet Saturn and to see the world beneath his microscope. He was the first to successfully apply the pendulum to the clock and he envisioned the use of a spiral spring to control a balance wheel in a clock.

Dr. Robert Hooke of England (1635–1703), inventor, mathematician and philosopher, gave us Hooke's Law, which says that the strain on a material is proportional to the force applied to it. He also stated the Theory of Inverse Squares. Hooke was the first to use the spiral hairspring to control the balance wheel in clocks and watches.

All of these eminent scientists were interested in instruments to measure time, an essential element in experimental research. Because of the basic research done by these and other scientists, clockmakers were able to enter a new phase in their craft: the quest for accuracy. This was a partnership with a common goal. The theory of the scientist, with the skilled hands and the conceptual powers of the clockmaker combined to give us precise timekeepers that we also used to measure the stars and the seas.

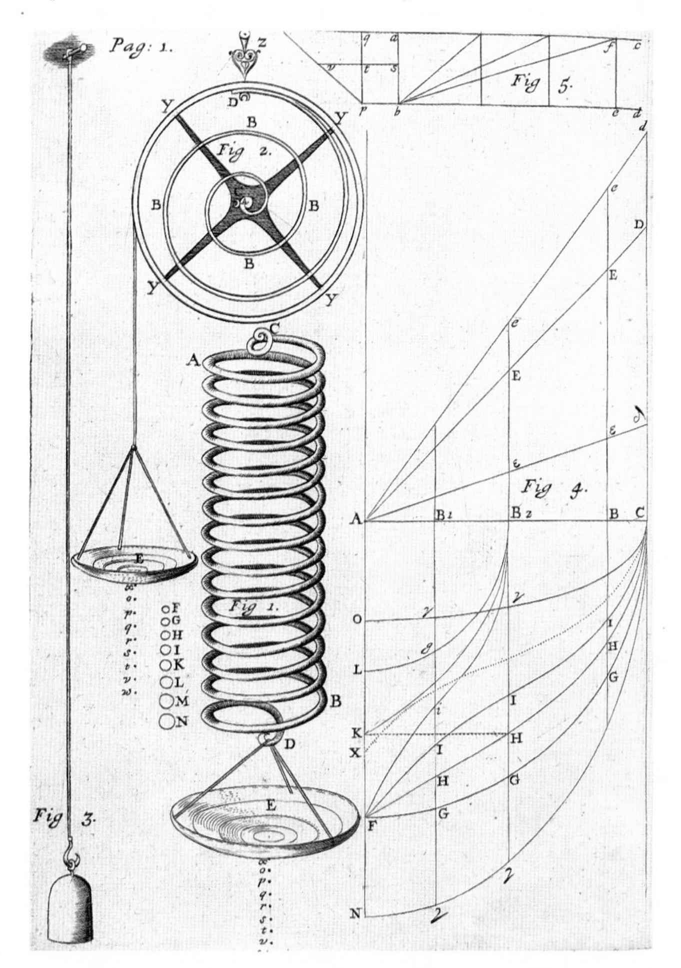

A diagramatic representation of Hooke's Spring, used to demonstrate Hooke's Law: The strain placed on the elastic body is directly proportional to the stress produced. In the diagram, the coil of wire (AB) is suspended from one end (C) and different weights are placed in the pan (E) hanging from the other end of the coil. This work led to Hooke's invention of the balance spring, which made the pocket watch possible.

The great Dutch mathematician and scientist, Christiaan Huygens (1629–95), made a study of the theory of the pendulum, which culminated in an applied pendulum designed by Huygens and made by Salomon Coster in 1657. Their success encouraged the manufacture of pendulum clocks in France and England, significantly increasing the accuracy of the clock to the point that the dial immediately became more prominent in order to reflect and exploit this new level of accuracy. Even earlier clocks were converted to pendulums to improve their accuracy.

The English long-case clock came about because of the pendulum. By enclosing the movement, pendulum and weights, the case protected these components from outside disturbance and at the same time made the clock a piece of furniture, adaptable to contemporary interior design and decoration. This fashion for designing clock cases to blend with the interior design and furniture spread to France, where it inspired a great revival in clockmaking. The clock mechanism determined the size and shape of English cases, but the French cases were developed into individual works of art, displaying all the elegance and grandeur of the Louis XV period.

THE QUEST FOR ACCURACY

The success of the pendulum clock immediately stimulated the quest for even greater accuracy. Science needed greater precision to study motion, acceleration and the positioning of places and objects on the earth, as well as for making astronomical observations. Scientists and clockmakers were ready to work together toward this end. They recognized the need for a stable time-frequency standard and a means of transmitting the power of clockwork to this standard.

The time-frequency standard is an oscillator giving a back-and-forth motion that chops up the power of the clock's gear train into measurable units that are then shown visually on the dial. Theoretically, the back-and-forth motions are equal on either side of their natural points of rest, but in practice there are variables that change this uniform oscillation. The challenge to clockmakers and scientists was to find the ideal oscillator—one which would maintain a constant

LEFT

Ahasuerus Fromanteel was a renowned London clockmaker in the mid-seventeenth century. He was the first English clockmaker to make pendulum clocks in collaboration with Christiaan Huygens of the Hague, who devised the first successful pendulum. This example of an eight-day long-case clock in ebonized pear-wood is one of Fromanteel's first pendulum clocks, circa 1665.

frequency—and refine the clockwork and its escapement to make this oscillating time-frequency absolutely uniform.

The verge escapement with its foliot oscillator was the first time-frequency standard for the mechanical clock. Making the erratic foliot (or pivoted lever) into a circular balance wheel did not enhance the accuracy of the clock until an elastic spring was added to help even out the oscillations. Replacing both of these oscillators with a pendulum increased accuracy dramatically. And then, around 1670, clockmakers replaced the crown-shaped escape wheel with a starlike escape wheel and an anchor escapement (see page 32), increasing the clock's accuracy even further.

Between 1715 and 1730, the designs and inventions of two English contemporaries, George Graham (1673–1751) and John Harrison (1693–1776), were the most influential in the development of the precision regulator clock. These clocks are also known as the *scientific regulator*, the *watchmaker's regulator* and the *astronomical regulator* clock. They recorded and maintained rates with up to one second per month in variation, setting a standard of accuracy unequalled until the end of the nineteenth and into the twentieth century.

England remained the leader in the development and manufacture of precision clocks through the eighteenth century and into the nineteenth centuries, although France and Austria were also significant contributors.

ABOVE LEFT
One of the oldest known pendulum clocks, dated 1657, built by Salomon Coster in Amsterdam, to Christiaan Huygens's design. Huygens was unaware of Galileo's pendulum theory of 1637 and had developed his own theory independently.

ABOVE RIGHT
George Graham, one of the most influential clockmakers, responsible for Britain's horological preeminence in the early part of the eighteenth century. Elected a Fellow of the Royal Society in 1721, on his death Parliament recognized his contribution with a burial in Westminster Abbey, next to his old mentor, Thomas Tompion.

THE CLOCKMAKER AND THE SCIENTIST

The quest for precision timekeeping attracted the most highly skilled craftsmen and the most learned men of science of the seventeenth and early eighteenth centuries in Europe. No single project had ever mobilized such talent. The cast of characters reads like a who's who in science: Galileo of Florence, Pascal of France, Hooke and Newton of England, Huygens of the Hague, Leibniz of Germany; and a who's who in clockmaking: in the 1600s, Salomon Coster in the Hague, Isaac Thuret in Paris, John Fromanteel and Thomas Tompion in London; in the 1700s, George Graham, John Harrison, John Arnold and Thomas Earnshaw in London; and Henry Sully, Pierre LeRoy and Ferdinand Berthoud in Paris. The quest: to measure the stars, the land and the seas, and the diverse mysteries in the laboratory.

Thomas Tompion (1639–1713), the father of English clockmaking and mentor to George Graham.

Salomon Coster (*d.* 1659) was the first to actually marry the pendulum theory of Galileo and Huygens to clockwork and prove the inherent accuracy of this oscillator.

Isaac Thuret (*d.* 1700) applied Huygens's balance spring to a watch and thus proved the inherent accuracy of the balance oscillator.

John Fromanteel (1638–80), working with mathematician Nicholas Mercator of Flanders, brought us the equation clock, built to show both solar time and mean time.

Robert Hooke (1635–1703) showed Thomas Tompion how to make an equation clock. And then **Julien LeRoy** (1686–1759) made a fine example of one. These clocks did not add much to the precision of timekeeping, but reflected the social changes going on at the time. Society, especially in the urban environment, was moving away from temporal time—time as marked by the sun—and moving toward a use of and dependence

A precision year-going regulator clock made for use at the Greenwich Royal Observatory in 1676. The clock originally had a pendulum that beat at two-second intervals.

A 1955 painting of Christiaan Huygens in Salomon Coster's clock shop, Amsterdam, examining their first pendulum clock—a device that revolutionized timekeeping accuracy.

upon mean time, or average solar time. It was beginning to divide the day into equal hours and months into equal days, leaving behind the natural, fluctuating daylight rhythms of ancient and medieval times.

George Graham (1673–1751), Tompion's employee, partner and then successor, invented the cylinder escapement, which succeeded the crown-verge escapement and added another dimension in increased accuracy. He also invented the dead-beat clock escapement used in long-case clocks and shelf clocks, which eliminated the recoil inherent in the anchor escapement and, in combination with his mercury expansion-compensated pendulum, was able to make astronomical regulator clocks that kept time to within a fraction of a second per day.

The others—Harrison (1693–1776), Arnold (1736–99), Earnshaw (1749–1829), Sully (1680–1728), LeRoy (1717–85) and Berthoud (1727–1807)—collaborated and competed to bring the world the most accurate portable clock: the seagoing marine chronometer.

Using their own inherent conceptual powers and the finest technical and handcraft skills known, the clockmakers of the seventeenth and eighteenth centuries combined the current science with their own surprising grasp of theoretical knowledge to push the limits of accurate timekeeping to the limits of the mechanical clock.

CHAPTER 3

FROM TABLETOP TO WAISTCOAT AND BEYOND

"Come what come may, Time and the hour run through the roughest day."

WILLIAM SHAKEPEARE

In the great clockmaking centers in Italy, France, Germany and England, the introduction of the coiled mainspring (see glossary, page 168) in the fifteenth century freed the clock from being a weight-driven wall instrument and made the table clock possible; with a handle it became portable enough to be taken on carriage trips. Further miniaturization made the portable clock small enough to be hung from the neck or from a belt. And thus the watch was born, some time in the late fifteenth century.

ABOVE
An oil painting by Haso da San Friano (1532–71) in the country where clockmaking probably began 200 years earlier. The portrait shows a wealthy man holding a miniaturized portable clock—a pocket watch—around 1558. The watch is spring-driven with an hour hand only and would have been one of the earliest watches made.

It was not an invention as such, but more the result of an evolution, a natural progression, and it evolved at several centers at nearly the same time and independently from each other. Initially clocks were made by blacksmiths and iron workers. But as the technology—with wheel works and escapement—evolved and brass and steel were used, clockmaking was picked up by locksmiths, who were accustomed to working in these materials and with small tools on a small scale.

THE WATCH-PRODUCTION SCENE

In the early sixteenth century, two distinct schools of watchmaking developed in France and Germany. The German makers devised a dumbbell-shaped oscillator powered by a mainspring whose strength was controlled by a counteracting spring called a *stackfreed*. The French, probably influenced by makers in Italy, used a wheel oscillator powered by a mainspring whose strength was controlled by a cone-shaped device called a *fusée* (see page 36).

Seventy-five years later, during the latter half of the sixteenth century, the watchmaking craft appeared in England with styles and techniques borrowed from Germany and France. For a hundred years, Germany dominated the watchmaking craft. This was a period of experimentation and trial and error. Watchmakers and apprentices traveled to distant watchmaking centers and returned to their home countries with new skills and knowledge. It was a period of a free exchange of ideas between makers in various centers in continental Europe and England.

The goldsmiths and jewelers of sixteenth century Geneva embraced John Calvin and his Protestant movement. However, when he declared jewelry and jeweled crosses and watch cases to be unacceptable personal adornments, the goldsmiths of Geneva traveled to Germany and France where they would be free to learn the art of watchmaking—a natural extension of the goldsmith's art. And thus another watchmaking center developed, in Geneva.

By the late sixteenth century the Geneva watchmakers had grown so numerous that they organized as a city guild to protect their members from competition and to control wages, as well as prices and the market.

ABOVE
The movement of a crown-verge watch with a stackfreed—a device for equalizing the power of a mainspring—and a dumb-bell balance oscillator to control the speed of the gear wheels as they revolve and turn the hour hand. Made by an unknown German maker in the late sixteenth century.

This was the era of craft guilds that existed throughout the European continent and in England. As watchmakers in any given area grew in number, they too sought guild status to prevent more watchmakers from setting up shop in their areas, thus protecting their limited market and their income.

By the year 1600 the Huguenots—Protestants in the solidly Catholic kingdom of France—were enjoying political and religious freedom under King Henry's 1598 Edict of Nantes. The number of Huguenot watchmakers increased, as did their production, so much so that by 1618 and the Thirty Years' War (1618–48), fought primarily in Germany, the German dominance was lost and for almost 100 years France would be the dominant watch center.

In the later part of the seventeenth century the wheel-cutting engine was invented and developed by watchmakers in the Prescot area of Lancashire. This engine had a revolutionary impact on the individual production of watch wheels: it not only increased output by an order of magnitude, it also ushered in the era of specialization in watchmaking. Rough movements (called *ébauches*, see page 115)—consisting of two plates with cocks, bridges, a mainspring barrel and

uncut *fusée* and gear train wheels—were made by specialist makers in small workshops situated in the area. The finishing process was achieved by a considerable number of workers, each with his own narrow skill, until the watch was completed under the watchful eye of the finishing watchmaker. The workmen would either be employed by him in his own workshop or factory or be "outworkers" in their own houses. This shift of manufacturing from individual and total handcraftsmanship to a reliance on outside specialists supplying parts to the finishing watchmaker spread throughout the watchmaking world to became the standard practice.

ABOVE
An early sixteenth-century German table clock. On the left is the gilt-brass clock case; on the right is the movement, showing a large balance wheel oscillator to control the rotating speed of the gear wheels and the hour hand that is attached to one of them. The maker of this piece is unknown.

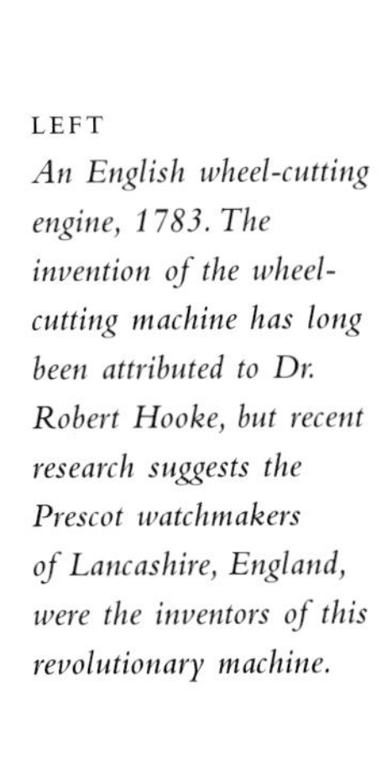

LEFT
An English wheel-cutting engine, 1783. The invention of the wheel-cutting machine has long been attributed to Dr. Robert Hooke, but recent research suggests the Prescot watchmakers of Lancashire, England, were the inventors of this revolutionary machine.

JOHN CALVIN'S INFLUENCE ON FASHION

John Calvin (1509–64) was one of the principal leaders of the Protestant Reformation of the Catholic Church. He was especially influential in Switzerland, England, Scotland and colonial North America. But it was not his religious teachings or his political views, of which he had many, that influenced watch fashion: it was his particular disdain for ornamentation, especially jewelry and the trappings of the aristocracy. His followers in France were called "Huguenots" and the English Protestants whom he influenced were called "Puritans."

John Calvin, French protestant reformer of the sixteenth century and predominant influence on the Geneva watchmakers and jewelers.

French by birth and education, Calvin settled in Switzerland and for the last 23 years of his life was the dominant personality in Geneva. In those days, Geneva was a tiny republic in a world of monarchies and principalities. It was probably the most creative and productive city in Europe. The Calvinistic Protestant movement attracted craftsmen, scientists, artists businessmen, and writers, offering them more freedom of thought and expression than the Catholic Church allowed.

When the Catholic Church in France began persecuting the Protestants, following the 1685 revocation of the Edict of Nantes, these victims fled to Geneva and London. Among the refugees were highly skilled watchmakers who brought their skills to both countries. They taught the now-blighted Geneva jewelers, who could not make jewelry any more because of the Calvin influence, the crafts of making watches and watch cases. They influenced the manufacture of quality timepieces, the designs of which were influenced by the austere dictates of John Calvin and his Protestant movement.

At the same time as the wheel-cutting engine came into being, three other events had revolutionary and significant impact on the craft. First, the Edict of Nantes was revoked by Louis XIV in 1685. This threw the French Protestant community into political and social turmoil. The edict had been instrumental in creating an environment of religious and cultural freedom that attracted craftsmen from all the known trades of the time: watchmakers, clockmakers, millers, textile workers, leather workers and many others. When the edict was revoked the Huguenots fled France for England and Geneva, whose various trades welcomed them for their skills and knowledge. Those who could not work in Geneva because of guild restrictions went into the nearby Jura Mountains to make component parts for the Geneva watchmakers. France's watchmaking community was decimated by this wholesale migration and immediately lost its world dominance in watchmaking to that of England and Geneva. France never regained its leadership, although it did produce some significant makers: LeRoy, Japy, Lépine and Breguet to name just a few.

The second revolutionary event around this same time was the invention of the spiral balance spring (see page 41) in 1675. Prior to this, the verge escapement—a miniaturized version of the clock escapement of the time—powered the balance wheel and the foliot. The spiral balance spring smoothed the oscillations of the balance wheel and eliminated its inconsistent ratcheting manner. With the introduction of the balance spring, the oscillator was returned after each oscillation in a more controlled and regulated manner. This increased the accuracy of timekeeping by several orders of magnitude and reduced the variance to minutes per day instead of hours per day.

Two men were responsible for this monumental invention during this time of great strides in science and mechanics. Dr. Robert Hooke of England and Christiaan Huygens of Holland, both working

BELOW

A 1590 engraving of Lake Geneva (Lac Leman) and its surroundings on the Swiss–French border between the Alps and the Jura Mountains. The City of Geneva, Switzerland, is on the shore of the lake and is bisected by the River Rhone.

GEORGE WASHINGTON AND LÉPINE

Such was the popularity of Lépine's design that George Washington, recently retired as president of the fledgling United States, sought out "a watch well executed in point of workmanship and about the size and kind of which was procured by Mr. [Thomas] Jefferson for Mr. [James] Madison, which was large and flat."

The watch he received through his emissary in Paris was from "Mr. L'Epine [who] is at the Head of his Profession here, and in Consequence asks more for his Work than any Body else. I therefore waited on Mr. L'Epine and agreed with him for two Watches exactly alike, one of which be for you and the other for me." (Gouverneur Morris in Paris, 23 February 1789.)

God. When they became portable, many of the clocks were in the shape of skulls, or crosses and crucifixes. to give the same message.

During the 1300s when rigid class systems were changing in Europe, the nobility began to dress far more elaborately to distinguish themselves from the newly emerging middle classes. Since it was only the aristocracy who could afford the technology of the clocks and portable timepieces, the cabinets and cases for these timepieces reflected the furniture and lifestyle of their wealthy owners.

Jewelers and goldsmiths in Geneva, then the world's jewelry center, created highly ornamented watch cases, encrusted with gemstones,

BELOW LEFT
A seventeenth-century Italian sundial made in the shape of a silver crucifix. Many watches made for the Catholic clergy bore this image as "memento mori" or a reminder of our mortality.

BELOW RIGHT
A silver-cased skull watch by J.C. Wolf, Germany, circa 1660. Timepieces of that period were often designed to remind users that every moment was precious because every tick brought them closer to their Maker.

ABOVE

A gold- and enamel-cased verge watch in the shape of a scent bottle. The maker is unknown, but it was probably made in France around 1640. By the seventeenth century, the influence of the French ornamenters on the art craftsmen was demonstrated by the making of "form watches"—watches made to look like animals, flowers, skulls, perfume bottles or anything but a watch!

for their wealthy clients. Their specialty was "form watches"—watch cases that looked like animals, flowers and skulls. They also made "cruciform watches" for the Catholic clergy and wealthy laymen of the Church; and they made "lunar watches" for the Muslim political and religious elite.

The timepiece as decoration found its highest expression in French watches in the late sixteenth and seventeenth centuries. Under the patronage of one of the richest courts in Europe, watchmakers went beyond highly engraved and jewel-encrusted enamels and created extravagantly colored paintings on enamel that were masterpieces in miniature art: exquisite cases, enameled and painted inside and out, that have never been matched since. By comparison, the Germans, Dutch and English preferred the highly engraved cases that were hand-chased, adding even more intricate detail to the design.

OPPOSITE
The earliest form of the gemstone-decorated watch case is the rock crystal case. Consisting of a hollowed out crystal base for the watch movement and a hinged crystal cover, the rock crystal case protected the movement, dial and hands like a metal case, yet allowed the user to see the time without opening the cover. This example is a nineteenth-century copy of a rock crystal seventeenth-century watch.

BELOW
The rock crystal case quickly led to a flat "window" of rock crystal held by tabs as shown in this star-shaped watch by Jacques Sermand of Geneva, circa 1630. The glass and plastic lenses of today are termed "crystals" because of this seventeenth-century influence.

THE INFLUENCE OF BUSINESS

As a culture of commerce grew and an urban style of life developed, fashion put a new spin on the clock. Just as time could be saved for religious growth and enlightenment, people, especially in the urban areas, began to realize that time could be saved for secular growth, too. Timepieces ceased to be reminders of death and became, instead, instruments of the living.

The "century of revolution"—beginning with the American Revolution of 1775–83, encompassing the French Revolution of 1789 and extending through to the series of popular uprisings in Europe known as the Revolutions of 1848—was a period of widespread revolt against these tyrannical, nobility-based governments and progress toward liberty and constitutional, or civil, government. The period manifested itself in fashion by substituting plainer attire for the elegant dress of nobility. And, since watches were worn on the person, they too were replaced by a plainer, simpler styling.Coincident with the changing political order, the changing social and economic order dictated the direction of timepiece development—the perfection of the clock and watch as an instrument of measurement. It was for the growing merchant and industrial societies of Western Europe that the clock and watch were depended upon to spur economic and social

ABOVE

An oval verge watch in a gilt-brass case made by Gylis van Ghule of London, dated 1589. The movement has a small balance and would have had a rather erratic motion as it was made before the balance spring was invented to even out the balance's oscillatory motion.

development. The astronomer recognized the precision timekeeper as essential for observation; the merchant for coordinating acquisition and distribution; and the manufacturer for optimizing production.

With the push for technical precision, adornment of the case was out, in a fashion sense. It seemed that too much ornamentation diminished the utilitarian value of the timepiece in most people's minds. Remember that adage: when technology stagnates, ornamentation flourishes? It seems that the reverse is true, also: when technological innovation flourishes, adornment suffers.

TIME ON YOUR HANDS

The first wristwatch was made for Queen Elizabeth I in 1571, but wristwatches, a further miniaturization of the pocket watch, did not become common until the late nineteenth century, and then only for women. This remained the custom until the early twentieth century when some men began sporting the wristwatch in Europe as a high fashion accessory; but, to most men, it was deemed far too effeminate.

It was World War I (1914–18) that brought about a complete change in viewpoint and started a fashion trend that rapidly became a necessity. Artillery officers, needing accurate time measurement to coordinate and launch their attacks, found that a watch strapped to their wrist made their work far easier. Pilots in the fledgling air war found the wristwatch far more convenient than the pocket or even the dashboard variety. U.S. soldiers, coming into the war late, found out how important time measurement was for modern combat: to synchronize troop movement and logistic support; to time field artillery, and to plan the arrival of coalition forces. Having the timepiece on the wrist freed both hands for the task at hand.

Military personnel returning to North America brought back a new appreciation for the wristwatch and a desire to have one of their own for their civilian work. From then on the wristwatch grew to be the timepiece of choice for men and women.

THINNER IS BETTER

There's an adage in fashion that states that thinner is better. And it seems nowhere truer than in the design of watches. From its beginning the driving force behind the technological advancement of the watch, along with precision, has been thinness. The pocket watch became thinner to fit into the waistcoat. Then it became thinner still in order to fit on the wrist.

But while fashion can drive the technology of the watch, the technology of the watch can drive its fashion, also. Once technology got the hand-wound wristwatch to near its ultimate thinness, a new wristwatch technology—the *self-winding watch*—demanded a thicker case to hold the additional mechanism that allowed the wearer's wrist movements to wind the watch automatically (in the late 1940s and into the 1960s). As the technology brought the size of the self-winding watch to near its ultimate thinness, the technology of

BELOW
A silver and enamel watch wound by the action of opening and closing the case. Made by Ebel S.A. of La Chaux-de-Fonds, Switzerland, circa 1930, it was worn suspended from a chain around the neck or in the pocket and represents a blend of fashion and technology at the time.

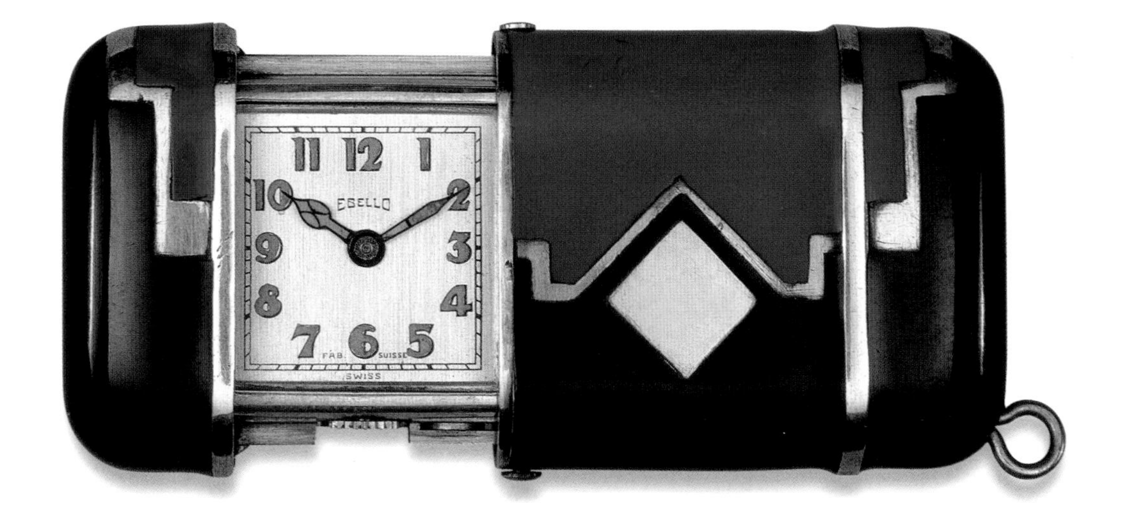

BELOW

A montre-à-tact *or "blind-man's watch," made of gold and enamel and highly decorated with turquoise and diamonds. A visually impaired person could tell the time by feeling the arrow (actually the watch's hand) and its relationship to the 12 turquoise hour marker beads. Made by Basil Charles LeRoy in Paris, circa 1810.*

the electronic watch required a thicker watch case to hold its battery and electronic circuitry. And then again, in the now familiar cycle, as the electronic watch neared its ultimate thinness, the new quartz watch technology of the 1970s required a thicker case to hold its technology. Today, quartz watch technology has gone so far that it can produce a watch so thin that just wearing it runs the risk of destroying its mechanism when the wrist is flexed.

FASHION DICTATES FASHION

As clothing fashion dictates simplicity in style, the watch reflects this simplicity. As clothing fashion uses more jewelry for accessories, the watch uses more gemstones—and begins to look more like a jewelry piece than a watch.

Today, the trend is for the watch to make a fashion statement, to reflect the wearer's lifestyle.

A thin watch for the business suit, a loose bracelet watch for women's business wear; sport-style watches for casual wear, the techno-look for the technology buff; the pocket watch for those yearning for a "simpler time" and reproductions of vintage-style timepieces for the nostalgic look.

Styling is in, and ornamentation is returning, not so much in the use of gems and jewels, but certainly in the proliferation of styles, both from the past and also from the designer's view of the future. Everything from simplicity of line to the audacious; from thin to bulky, with cases and bands available in every color of the rainbow—and then some: it's all acceptable.

There are innovations occurring in watch-case materials, manufacturing and styling, but no great strides in watch technology itself. We must be at the end of another one of those periods in which technology stagnates and adornment flourishes.

ENAMELING

For much of the seventeenth century the most important watch-production center was Blois, France. When the Edict of Nantes was revoked in 1685, however, (see page 48), many watchmakers, enamel painters and goldsmiths fled France for England and Switzerland.

One of the individuals of note who fled to England was Jean Toutin (1578–1644) an enameler who developed a new technique for small objects by painting a picture using a thin brush and a thin coat of paint onto a layer of previously fired enamel. The object was then fired again, giving it an incomparably transparent quality. Toutin painted portraits of rulers and nobility on items not much larger than an inch and a quarter square. His technique was immediately applied to small watch cases.

Pierre Huaud (1612–80), a refugee Huguenot, and his three sons moved to Geneva in 1630; and he was granted Swiss citizenship in 1671. Even today, the artistic talent of the Huaud family with their beautiful, delicately painted portraits and landscapes on pocket watches can be seen in museums and private collections. Not surprisingly, when the Edict of Nantes was revoked many French enamelers followed the Huaud family to Geneva.

A late sixteenth- to early seventeenth-century engraving by Hagenberg, depicting the massacre of the Huguenots. The house of Cabrière in Cabors, France, is being torched by Catholics, November, 1561.

CHAPTER 4

THE CRAFT ERA IN WATCH-MAKING

"Of all, clockmakers and morticians should bear the keenest sense of priority—their lives daily spent in observance of the unflagging procession of time ... and the end thereof."

DAVID PERKIN'S DIARY, JANUARY 3, 1901.

ABOVE
A blacksmith in his forge—a French painting by Antoine Lenain (1585–1648). Initially clocks were made by blacksmiths and ironworkers but as the technology of clocks evolved to include brass and steel, clockmaking was picked up by locksmiths and goldsmiths, who were accustomed to working with these materials on a smaller scale.

The real story of watchmaking, as distinct from clockmaking, was concentrated in a very small area for the first 300 years of its 500-year history, evolving from central Italy in the fifteenth century to France, Germany and then to southern England in the seventeenth century. Around the turn of the eighteenth century, watchmaking expanded to western Spain, northern England, Scotland and Wales, and to the lower reaches of Scandinavia and east to St. Petersburg.

THE WATCH

No one knows for sure where the word *watch* came from, though it's thought to have been once related to the word *wake*, in the sense of a vigil. Presumably, some of the first watches were used during the 1500s by town watchmen, or guards, in Europe. A watchman carried a portable clock around his neck or waist to time his rounds. These rounds were called *watches*, and supposedly the small portable clock associated with these rounds became known as a watch, also. Or perhaps it was because a portable clock hanging around the neck was so close to him all the time that a person tended to "watch" his time as he went about his business and ordered his life to the clock.

Miniaturization allowed portability and as soon as it became fashionable to wear a watch, watchmakers sought further miniaturization. The blacksmiths, from whose ranks clockmakers arose, did not have the skills for tiny work. Such miniature work called for the eye–hand skills of the locksmith, goldsmith and jeweler. And it was from these ranks that the watchmaker evolved.

PASSING ON THE CRAFT

The specialized skills needed to make the intricate wheels and levers and springs of the watch took a long time to learn. And, once learned, they took a lot of practice to perfect. Makers recognized the close link among the skilled hands, focused mind and creative thought needed to develop a novice into a master.

In the mid-sixteenth century watchmakers inherited the ideal of the consummate craftsman from the other established craft trades: a good workman should be able to make a complete watch from start to finish. To this end, a novice typically indentured himself to a master watchmaker as an apprentice. Ideally, the apprentice would sit at the master's feet and soak up all his knowledge and skill. In reality, the master had work to turn out in order to make a living and he was not going to have some ham-fisted adolescent touch these precious, precision instruments that he had been laboring on. So it turned out that the apprentice, typically 12 to 15 years of age, would sweep the floor, carry in wood and coal, stoke the fires and hammer brass into usable sheets until the master got accustomed to his presence and learned whether the two of them could get along together in an intense, long-term relationship, with the young man living in the master's home (or, more typically, in the shop) and being fed by the master's wife.

ABOVE
The timepiece as decoration found its highest expression in the French watches of the late sixteenth and seventeenth centuries. Under the auspices of the richest courts in Europe, the watchmakers of Blois and Paris used polychrome paintings on enamel that were masterpieces of the miniaturist's art, as in this French timepiece of the late eighteenth century.

In between the menial tasks of sweeping and stoking, the apprentice would learn to file, practicing with scraps of brass and steel for months and months on end: endless drudgery for ten or twelve hours a day, six days a week. Learning to saw with the fine piercing saw took more months of tedious drudgery. But all the time the apprentice was developing the critical skills of eye-hand coordination and cultivating a soft, gentle touch and feel for the tiny, delicate parts. He would learn also to manipulate fingers and parts in distances measured in millimeters and fractions of a millimeter, and had to learn patience and develop a critical eye for detail. Acquiring *that* degree of perfection was the only acceptable measure of one's accomplishment.

Two years into the apprenticeship, the student would have developed enough patience, tolerance and lightness of touch to operate the hand crank or foot pedal for the lathe used by the master. Thus the apprentice would work silently alongside the master, providing a source of power as the master brought the hand-controlled cutters to bear on brass and steel parts—turning out wheel blanks, plates and arbors for the watch he was building, boring recesses into plates and drilling precisely located holes.

The apprentice was by now becoming useful, instead of a burden. He would slowly acquire the skills, the patience and the focus, and in time he would develop the creativity needed to design, lay out and execute a watch as his master did. Typically, it took five to seven years, six days a week, 52 weeks a year to complete the apprenticeship. Sundays were reserved for church and recreation.

The apprenticeship ended when the apprentice presented his master with a masterpiece: a watch made from start to finish with his own

BELOW
A French necklace watch, circa 1680, displays enameling in the style of the Huaud brothers. The back of this watch shows the "Caritas Romanas" (Roman Charity). The Athenian woman nurses her father, who is wasting away in prison, with her own milk.

5.
HOROLOGIA FERREA.
Rota æqua ferrea ætherisq; voluitur, Recludit æquè et hæc et illa tempora.

ABOVE
A seventeenth-century clockmaker's shop with the master directing the activities of the journeymen and apprentices under his charge. Typical activities included everything from forging their own iron and brass to handcrafting the components and assembling and adjusting the completed timepiece.

hands. And then it would be another year or two working as a journeyman watchmaker in his master's shop, actively receiving wages for his labor, before he would be released to work for another shop or, if he had the funding, to open his own.

This was the way in which skill and knowledge were passed from one generation to the next. The master was the shop owner, the employer and the teacher; the apprentice was the student; the journeyman was the skilled employee-craftsman working for a wage and looking forward to the day when he could open his own shop and become a master himself. Some remained employees, working with the same master, or journeyed to other shops, depending on employment opportunities, their own perceived worth or what they thought might be better working conditions. A gifted journeyman would marry the boss's daughter, eventually inheriting the shop; or even marry the boss's widow. In a society in which women were generally excluded from ownership or paid employment, it was a system that provided for the succession of a watchmaker's shop, his skills and his knowledge from one generation to the next, if there was not a son to do the inheriting.

CRAFT GUILDS ...

From ancient times, crafts were organized into specialties, based on the raw materials that they used. Metalworkers formed one trade, millers another, textile workers another and leather workers still another. As communities grew and political structures came into place, these various trades organized into legal corporations or guilds to protect their trades and regulate their industries.

Clockmaking, when it first became a profession, was included in the metalworkers' guild. When the growth of the clock industry in a

LEFT
A spring-driven tabernacle-type table clock by Samuel Haug of Augsburg, circa 1630. The tower or tabernacle clock takes an architectural form so that the complete clock appears to have a column at each corner. These clocks have their bell at the top, an alarm mechanism and a striking train that counts the hours.

THE GENEVA RULES OF 1601 (ABRIDGED)

Rules and regulations governing the Corporation of Watchmakers reviewed and approved in council, 19 January 1601. [Note: each article in these rules had a fine or fee associated with violations or applications, which are not included here.]

I All the assembled master watchmakers will be obligated to pray to God, asking His Presence within their midst, so that they would only say or do only those things that would honor God and benefit the City.

II Two masters will be chosen to govern, inspect and oversee the Corporation so that only good work and honest merchandise would be made.

III No one may take more than one apprentice and for no less than 5 years; or 2½ years if he is a journeyman armorer or locksmith. The master may take a second apprentice at the end of the 3 years of his first. The apprentice will pay a fee, half of which goes to the commissioners and half for the maintenance of the Corporation and/or for poor itinerant journeymen.

IV No apprentice may break his commitment to his master. If he does so, he must start his apprenticeship all over again when he is returned to his master.

V No apprentice can ask to be a master until he completes a full year as a journeyman.

VI To open a shop and become a master, a journeyman must make two masterpieces: a small watch with a wakeup alarm to wear around the neck; and a square table clock of two heights; both to be reviewed by all masters of the Corporation and upon payment of an examination fee to the Corporation.

VII To apply to make the masterpiece, the apprentice must have the recommendation of his master.

VIII No master may lure a journeyman away from his current master by any means.

IX The master is free to take another apprentice if an absent apprentice does not return after 2 or 3 months' absence (except in the case of illness).

X No master can buy any watch work either begun or completed by a journeyman or apprentice. If he is offered any to buy, he must report the offer to the "master juror" of the guild.

XI No master from outside the city can set up shop without being passed as a master or presenting proof of having passed as a master in another guild.

XII All sons of masters must make a masterpiece watch before being allowed to open their own shop.

XIII No master is allowed to sell a piece of work given to him for repair. Upon a second offense, he will lose his mastership.

XIV All the "master jurors" (guild inspectors) can visit any shop at their pleasure, to inspect for good and honest work. They can break substandard work and must present the substandard work to the commissioners of the guild.

XV Merchants who do not belong to the Corporation are forbidden to sell or trade watches or clocks within the City.

XVI A master who refuses to serve the Corporation in his elected capacity will be fined.

XVII Every master must sign his work.

XVIII All present masters must swear to uphold the articles of the Corporation so that everything may proceed to the honor of God, the profit of the City and the preservation of this Corporation of watchmakers.

community reached a point where members' commercial and professional interests differed materially from those of the other metalworkers, the clockmakers would organize as an independent guild. The clockmaking craft broke down into three categories, just as did all of the other urban crafts: the master, who owned the shop and equipment and who was the employer and teacher; the student, who was the apprentice; and the journeyman, who had completed his apprenticeship but was not yet in a financial or political position to open his own shop.

This division existed long before craft guilds came into being and long after they were gone. The guilds merely legalized and institutionalized this hierarchy, for a while.

During the sixteenth and seventeenth centuries the number of craftsmen manufacturing clocks was sufficient to establish specialized craft guilds in most of the clock-production centers in Europe. Augsburg and Nuremberg were two of the most important of these centers in the early days of the guilds.

Usually, watchmakers were included with clockmakers, being differentiated only by the size of the work that they did: large work, clocks, and small work, watches. The guilds were intended to maintain a high standard of craftmanship and to protect the interests of their members, which they did. The guilds protected the craft's physical territory and their place in society. They were fraternities that regulated entrance into the craft, balanced the number of makers in an area so that all could earn a living, prohibited other crafts from making and selling what had been reserved for them, and prohibited the importation of competing products, even by their own members. In other words, the guild or corporation of craftsmen monitored and ensured the survival and prosperity of each maker's shop, even to the extent of seeing to it that the less efficient, yet still talented, makers got their fair share as well.

The guild certainly ensured the transmission of quality workmanship, and it made possible the production of extraordinarily complex and artistic products, products that would take years to complete and would not even have been

BELOW

French oignon *watch with a crown–verge escapement and a pierced and engraved case, circa 1700. The word* oignon *(the French word for "onion") is a colloquial name applied to the large French watches of the late seventeenth and early eighteenth centuries.*

RIGHT
A disassembled seventeenth-century crown-verge watch movement showing the major component parts. (top) The inside of the upper movement plate with the crown wheel; (first row, left) fusée; (center) balance with the verge staff; (right) balance bridge; (second row, left) mainspring barrel; (center) contrate wheel; (bottom) lower movement plate.

individual components. One would make wheels, another would make arbors and yet another would make the plates upon which all the other components would be supported. Spring makers, escapement makers, jewel makers, balance-wheel makers—all were specialists making components for the watchmakers. In his shop the watchmaker and his employees would file and fit, articulate, and adjust the whole assembly into a complete watch. The watchmaker would design the movement, even make a prototype and contract with each specialist for the needed components, often financing their setup and production, in order to have the needed supply.

Nowhere was the system more exploited than in the Jura Mountains of Switzerland. Here the cottage industry began in the first quarter of the eighteenth century and was a natural for these mountain people, locked into their homes by the winter snows in a land of little natural resource but with a people of great natural skill, ingenuity and motivation.

In London, Coventry, Clerkenwell and Lancashire, and in Geneva, Paris and other watchmaking centers, similar systems of component manufacture developed. But do not let the term "manufacture" imply mechanization at this point. The components were still handmade (or nearly so) on simple hand-cranked machines. The individual components were hand-fitted and hand-assembled by journeymen who had the skills to make a complete watch, but who for expedience adopted this method of finishing watches. The master's jobs were to design, lay out, contract and bring all the components into one place for fitting and assembly, and to oversee the quality of production.

This method led the way to the *ébauche* system of watchmaking. Instead of the components going to a finishing watchmaker, they went to a sub-assembler (the *ébauche* maker) who fitted the wheels and plates together, making sure they were uprighted in the plates and turning freely. This *ébauche*, or rough movement, was then sold to a finishing watchmaker who added the balance wheel, escapement, mainspring, dial and hands, finished the plates, jeweled the wheel-arbor pivot points, and adjusted the watch to his individual standard of perfection.

This was the state of watchmaking at the height of the handcraft era, an era that had lasted for 300 years, from the 1500s through into the 1800s. The system of manufacture had evolved from the total crafting by an individual, to the handcrafting of component parts by specialists. The hand skills, knowledge and craftsmanship were passed from master to apprentice throughout these three centuries, protected for a while by a guild system, but transcending it to the brink of a new era at the turn of the nineteenth century. The evolution of watchmaking was poised for further development, this time as the mechanization of the watchmaker's art.

BELOW

A women's 9-carat gold quartz wristwatch with a gold bracelet in the "Greek key" pattern. High quality watches such as this require the unique talents of the modern watch repairman to maintain their beauty and accurate usefulness.

THE CRAFT ERA CONTINUES

Yet, as the watchmaker's art evolved into the machine age with its mass manufacture of watch movements, and then took a dramatic turn away from mechanical watches and into electronic watches in the 1960s and 1970s, the handcraft principles and skills continued in the fields of antique restoration, repair and the making of complicated watch mechanisms. Formal watchmaking schools came about in the mid-nineteenth century in most of the industrialized countries of the world and continued to teach many of the handcraft principles and skills. The schools augmented the apprentice system and later on replaced the apprentice system entirely. Even today, those who want to learn the finer points of antique watch restoration can link themselves with a knowledgable watchmaker who becomes a mentor. And in the spirit of the apprentice system, mentors can continue to pass their knowledge on to their apprentices, one generation after another.

THE CRAFT ERA IN THE QUARTZ AGE

The watchmaking industry took a beating during the early 1960s when the electronic watch emerged, followed by the quartz watch a decade later. Consumers abandoned their old mechanicals for the new technology in keeping time.

With the maturing of the quartz watch and dramatic reductions in the cost of producing these watch movements, the vast majority of quartz watches became a throwaway commodity. It became almost a break-even proposition whether to buy a new quartz watch or service the existing one.

For generations, good-quality watches had been given as gifts or as symbols of significant accomplishments. With a modicum of care and service, they could be expected to last for generations, even passed down to the next generation as family heirlooms and mementos. By the mid 1980s many of the new watches that were being received as gifts and mementos were no longer treasured. In the minds of many, their lower prices demeaned their value as a gift. Although they were extraordinarily accurate and dependable, they had become just an appliance.

The Swiss, especially, put quartz movements into high-value cases in response to consumer demand for a luxury watch that would have lasting value, not only for personal use but for gift-giving, too. Electronic watch movements have a finite lifetime. After so long, even if serviced often and faithfully, the electronic components deteriorate and break down. Quartz movements have become so standardized that those deteriorated movements can be replaced with new ones that fit the case and dial configurations. But there is still that throwaway aspect, even with the fine cases of the luxury brand watches.

OPPOSITE
Gold and leather triple-cased watch by Thomas Mudge. The watch has a cylinder escapement, and a minute repeating mechanism that sounds the current time to the nearest minute on a gong. The watch was made in London, circa 1755.

RE-ENTER THE MECHANICAL

There's something about the mechanical watch: its long history implies lasting quality, serviceability and value. Its steady rhythmic tick gives life to a seemingly inanimate machine. Consumers picked up on this in the mid 1980s after 25 years of the electronic watch. Vintage watches—mechanical watches of the 1940s to 1960s, as distinct from antique ones—saw dramatic sales increases on the used watch market

ABRAHAM-LOUIS BREGUET

A-L. Breguet, watchmaker extraordinaire (1747–1823).

Arguably the greatest, most innovative watchmaker who has ever lived, Abraham-Louis Breguet epitomized the height of the handcraft era in watchmaking.

From the early 1780s onward, Breguet led the way in almost every branch of watch- and clockmaking. Once he opened his shop and assembled his team of craftsmen, Breguet himself did little of the actual assembly of watches or clocks. He designed them. His creative mind created the unique, the innovative and the highest of quality. The crowned heads of Europe and the wealthiest of merchants waited months and even years for the watch or clock of their dreams and specifications. He held his shop to the highest of standards. So fine were his creations that even today those few who can afford them treasure them with great pride and joy.

After Breguet, watches were, at best, copies of his designs and conceptions. From 1776 to 1823 (a span of nearly 50 years), Breguet's shop turned out 5,000-plus watches made by his own workmen to his exacting standards. Breguet modified the newly invented Lépine movement layout to his purposes (the Lépine movement was a watch movement design; the Lépine style refered to a thin watch case without a metal cover over its face): and set the standard and style for the pocket watch, inside and out—movement, case, dial, hands as well as quality—far into the future. Breguet relied upon Geneva makers to supply his rough movements (*ébauches*) so that his artists (as he so unabashedly called his highly skilled craftsmen) could keep up with demand.

From his cheapest watch—the "subscription watch," batch-made and paid for in advance (they cost a French laborer a year's salary)—to his self-winding pocket watch; from his highly precise ship's chronometer to a clock that wound and reset your pocket watch while you slept (the *sympatique*); from his addition of a second hand to the pocket watch dial to his shock protection system, Breguet set standards in innovation, style and technique that have yet to be equaled.

Breguet had a great reputation, developed from his earlier years as a superb watchmaker, but he proved to all that it takes more than a great name: it takes quality, efficient techniques, low costs and access to large markets to make an industry.

The single hand Breguet* souscription *watch, Paris, 1798. The watch was so named because purchasers ordered the watch by subscription.

and in auction houses. People began wearing them, not just collecting them. Sales of new mechanical watches soared in the 1990s as Swiss makers romanticized the aura around the mechanical watch.

From the vintage mechanical watch to the mass-produced ones of today, to one-of-a-kind specialty watches, the mechanical watch is enjoying a real resurgence in interest and popularity. A number of very talented and skilled watch craftsmen are finding a new market for their unique craft-era skills. Leaving the repair and restoration field, they have moved into finishing and enhancing new mechanical *ébauches*. The addition of unique specialized escapements, automatic winding mechanisms, perpetual calendars and other interesting features makes many of these watches rise above the ordinary. Some contemporary craftsmen have created their own *ébauches* and then added their own features, combining the handcraft methods derived over the previous 450 years of watchmaking with the machine tools and manufacturing techniques of today.

BELOW
The modern Omega men's wristwatch. This long-established firm, working in the tradition of the établisseur, *brings components from various manufacturers to their own factories in Switzerland to complete the watches they acquire from* ébauche *manufacturers. Omega was founded by Louis Brandt (1825–79) in 1848 at La Chaux-de-Fonds, where they made precision key wind pocket watches.*

Some watchmaking firms, both new ones and long-established ones working in the tradition of the *établisseur,* are bringing components from various suppliers into small factories and employing talented watchmakers to finish and enhance the *ébauches*, and are now producing thousands of truly handcrafted mechanical watches (see page 115). Some make only a few pieces, augmenting their repair and restoration profession and fulfilling a desire to create from scratch an example of what they've been spending a career restoring. And a very few others, in the tradition of the early watchmakers, are making just a select few totally handmade watches each year for those very wealthy individuals who can appreciate and enjoy these original and contemporary works of art.

THE NEW RENAISSANCE WATCHMAKER

One watchmaker of today has led the handcraft era within the age of the quartz watch. George Daniels has been called the greatest living watchmaker and a "renaissance watchmaker" because he has shown a diverse interest and expertise in all areas of traditional watchmaking, from the mechanical to the artistic, from the scientific to the inventive.

ABOVE
The Omega De Ville Co-Axial gentleman's wristwatch, a self-winding (automatic) mechanical wristwatch using George Daniel's co-axial escapement, circa 2001. The co-axial escapement is one of the twentieth century's major horological inventions—eliminating nearly all friction, a major source of lubrication problem in the mechanical watch, and giving an accuracy approaching that of the electronic quartz crystal.

Through his work, his publications and his lectures he has led a revival of the classical art of watchmaking beginning in the United Kingdom and then Europe and even spreading to North America.

Born in the East End of London in 1926, George Daniels started his illustrious career as a professional watchmaker at the age of 22. His intuitive skill, fertile mind and total absorption in the art and craft of watchmaking drew him to the watch restoration field as his skills and talents advanced. But restoring the fine works of others left him yearning to create his own masterpieces. By the time he was 43, Daniels had made his first watch. Over the ensuing 26 years he made a total of 27 watches, all completely handmade, including the cases, dials and hands. All were complex watches to some degree, the least complication being a revolving *tourbillon* (see page 171) cage to carry a chronometer escapement and balance wheel in one-minute revolutions, offsetting the effects of carrying the watch in various flat and suspended positions. Many others displayed various complex mechanisms, among them sidereal time, equation of time, and chronograph functions with lapsed-minute recorders, minute repeaters, and winding indicators.

In 1981 Daniels created his co-axial escapement (see page 79), a unique watch escapement that was unaffected by the viscosity of oil and would perform with an accuracy normally claimed only by the quartz watch. Over the following 15 years he incorporated this escapement into the next eight watches that he made, refining and testing its performance as he went. He miniaturized and further refined the escapement to fit into a wristwatch's movement, putting it into eight high-quality Swiss wristwatches, all of which were worn for long periods of time to serve as test vehicles for this revolutionary new escapement design.

Daniels's intent from the beginning was to create an escapement that would allow the mechanical watch to perform with the accuracy of a quartz watch. His express intention was to revive an interest in the mechanical watch in the consumer's mind. Fifteen years of rigorous testing proved to Daniels that his co-axial escapement was the only serious contender to the lever escapement in modern watches. Yet persuading major watch producers to change from the 200-year tradition of the lever escapement and put this new escapement into commercial production proved to be a daunting task.

But persuade them he did. In 1999 the Omega Watch Company announced that it had developed a watch with the Daniels Co-axial Escapement, bringing an end to a nearly 30-year personal campaign to develop a new escapement and see it used in volume commercial production. It took a further three years of refinement by Daniels and Omega's production engineers before they were able to make mass-production of the escapement commercially viable.

Instead of the lever escapement sliding on the escape-wheel teeth as it impulses the balance wheel back and forth, the Daniels escapement flicks the escape-wheel teeth, touching only momentarily. Because of the reduced friction, this new escapement doesn't need lubrication; and because it doesn't need lubrication it needs to be cleaned only once every ten years or so. Less friction also means better timekeeping because the impulses transmitted to the balance wheel are much steadier—just as John Harrison proved 250 years ago (see pages 102 and 132), and just as Thomas Earnshaw and John Arnold successfully demonstrated with their portable marine clocks, which are still made to their designs 200 years later.

Daniels says that, if the watch is properly regulated to the user's wearing habits, it will be as accurate as a quartz watch, running within three seconds or so a month.

BELOW
The escapement consists of an intermediary wheel A, a double co-axial wheel B, consisting of pinion C and escapement wheel D, lever E with 3 ruby pallet-stones F, G and H, and a roller I with a ruby impulse stone J, and an impulse pin K. The clockwise impulse is provided by the teeth on escape wheel D directly engaging the ruby impulse stone J. Counterclockwise impulse is provided by the teeth of escape pinion C engaging the lever impluse stone G. After each impulse the escape wheel is locked stationary by the pallets F and H, allowing the balance to complete its vibration.

THE DANIELS CO-AXIAL ESCAPEMENT

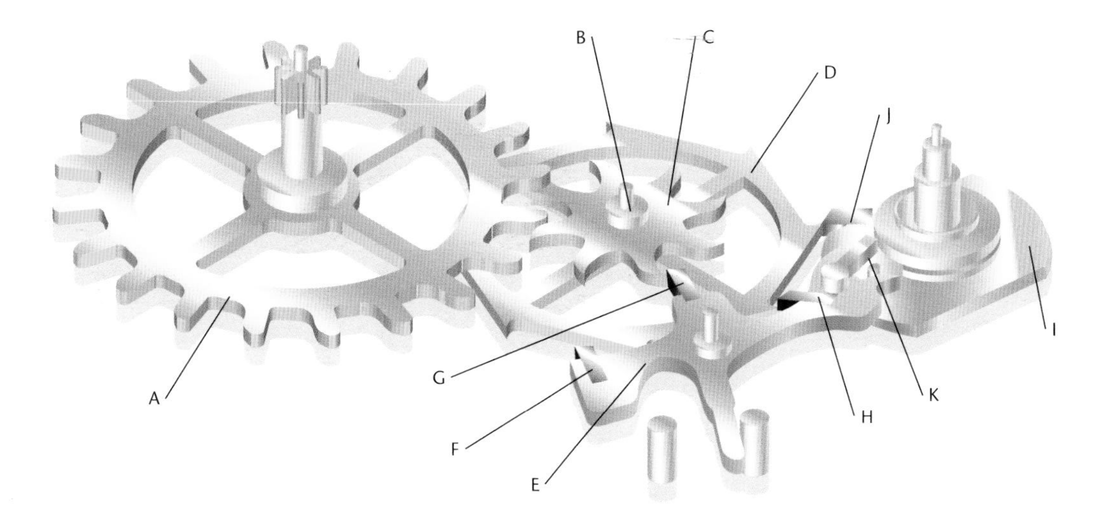

THE QUEST FOR THE PRECISION WATCH ESCAPEMENT

The search for the perfect timepiece escapement has spanned the years of watch- and clockmaking. There have been hundreds of different types of escapements created over the nearly 700 years of timekeeping history, but most were produced as only a single prototype or in a very limited series. Others had longer lives but eventually fell out of use because of poor performance or difficulties in mass-production.

But only five escapements have been used for any length of time at all. The first one was

THE CYLINDER ESCAPEMENT

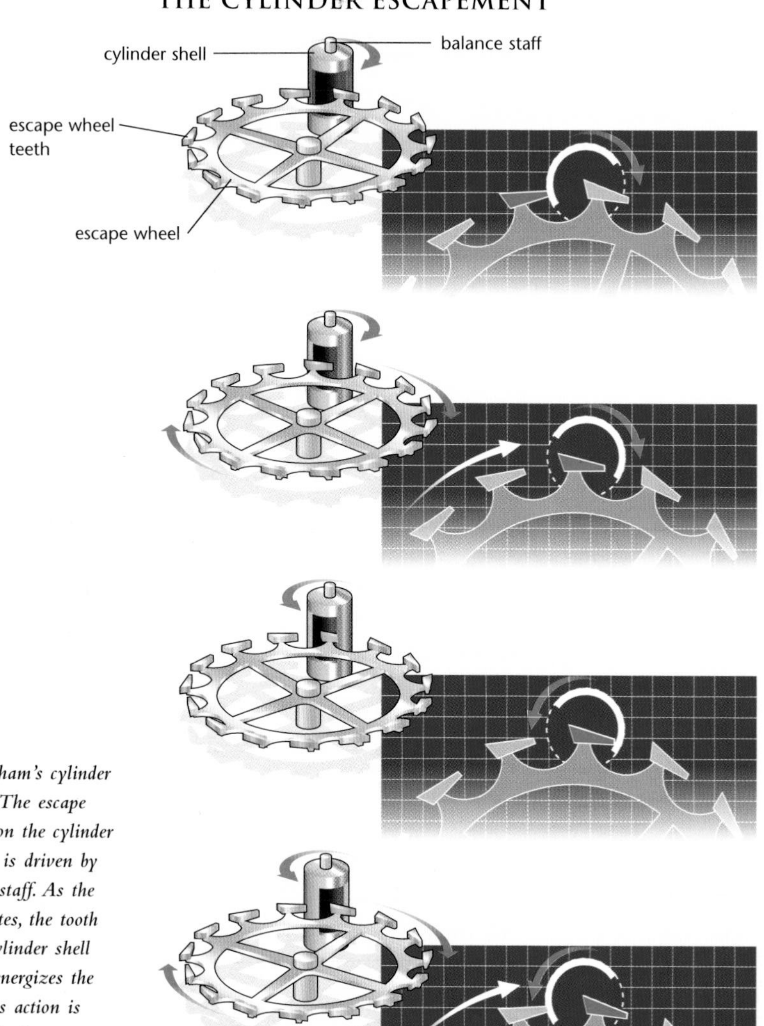

George Graham's cylinder escapement. The escape tooth locks on the cylinder shell, which is driven by the balance staff. As the balance rotates, the tooth enters the cylinder shell cavity and energizes the balance. This action is repeated with the next oscillation that drives the gear train.

the *verge escapement* (or crown-verge) that was used continually from the first clock in the early 1300s until the middle of the nineteenth century. In about 1720 the cylinder escapement, invented by George Graham, was used by some of the finest watchmakers in England for about 50 years. It was the preferred escapement of the gifted French chronometer maker Ferdinand Berthoud (1727–1807) and the watchmaker master, Abraham-Louis Breguet (1747–1823). The cylinder escapement survived into the 1950s, albeit as a modified, low-cost escapement for low-priced watch movements. At about the same time as the cylinder escapement, the *duplex escapement* was used, also in England.

Thomas Mudge, the inventor of the detached lever escapement, London.

The duplex proved a more accurate escapement than the cylinder but it was more delicate. But all of these three escapements had one major disadvantage: The escape wheel was almost constantly in contact with the balance wheel. This continually disturbed the rhythmic oscillation of the balance. Watchmakers came to realize that the balance wheel had to be able to oscillate freely if it was to maintain any kind of uniform frequency rate.

To give this desired oscillation, two detached escapements were developed in the mid-eighteenth century, replacing the previous three, and were used almost exclusively for the next 250 years, right up until today.

THE DETACHED LEVER ESCAPEMENT

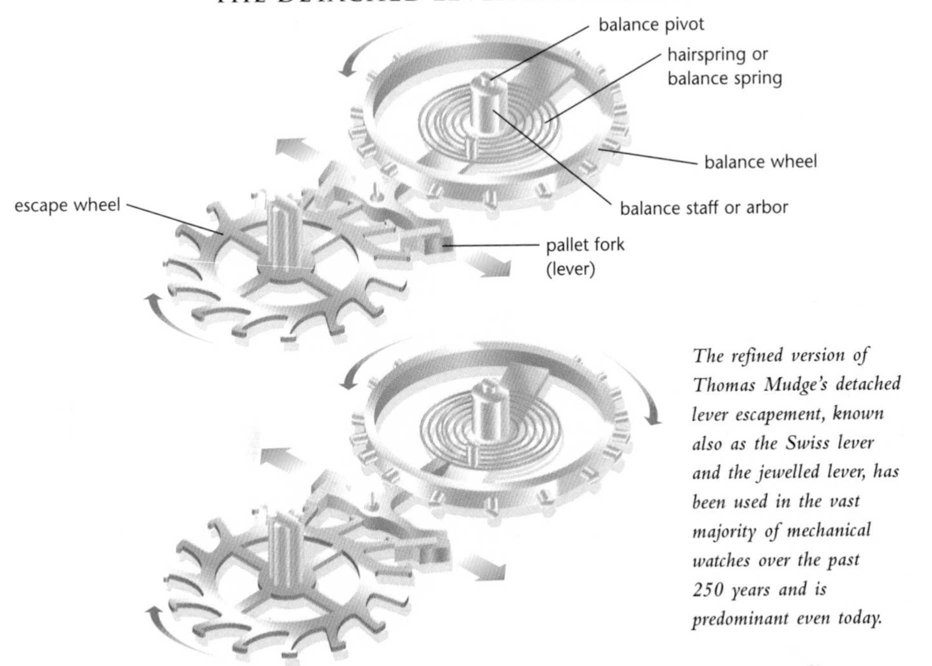

The refined version of Thomas Mudge's detached lever escapement, known also as the Swiss lever and the jewelled lever, has been used in the vast majority of mechanical watches over the past 250 years and is predominant even today.

DETACHED ESCAPEMENTS

Around 1760 two watchmakers, one in France and one in England, created detached escapements, i.e., escapements whose escape wheel came into contact with the balance wheel only long enough to energize the balance and keep it oscillating. The detached escapements thus invented were in two forms: the *detached lever* by the Englishman Thomas Mudge (1715–94); and the *detent* escapement by the Frenchman Pierre LeRoy (1717–85).

DETACHED LEVER

Mudge's lever escapement used a Y-shaped lever called a *pallet fork* that—powered by the gear train and a spring—locked and unlocked the escapement, rocking the tail of the fork back and forth. At each angular movement of this rocking fork, the tail would catch the balance wheel and send it on its excursion, immediately unlocking and staying out of the way until the balance returned for another energizing impulse. The role of the lever was to convert the linear motion of the gear train into the angular, oscillatory motion of the balance wheel. Mudge's lever escapement was refined over the next 50 years or so by other watchmakers until it became accepted as the escapement for a quality watch, a position it still holds today.

A high-precision gold-cased pocket chronometer made by Sylvan Mairet of the firm of Hunt & Roskell, London, 1836. The watch has separate subdials for the hour (right), seconds (left) and a center minute hand for easier, more accurate reading.

DETENT

LeRoy's detached escapement was a marriage of Mudge's lever and the *detent* escapement that was to evolve in the hands of John Arnold (1736–99) and Thomas Earnshaw (1749–1829), both of London. The *detent* escapement is locked by a spring just after the wheel gives an impulse to the balance wheel, leaving the balance wheel free to oscillate until it returns for its next energizing impulse.

The *lever* escapement is very stable, maintaining its action even when disturbed by outside forces. It impulses the balance wheel in both directions, back and forth, and it is self-starting after its mainspring is wound from a stopped position. But it needs lubrication to transmit energy to the balance wheel efficiently. The *detent* escapement is sensitive to outside disturbances. It impulses the balance wheel in only one direction, letting the balance wheel oscillate back and forth without disturbance; and it is not self-starting, the balance wheel needing to be rotated physically from one side to the other to restore its motion after its mainspring has wound down and the watch has stopped. However, the big advantage of the *detent* escapement is that it does not need lubrication. It is the most efficient of all escapements, having been used in marine chronometers (sea clocks) and a number of different watches since its invention in about 1785 right on up until today. However, it was yet to be perfected...

THE LEVER/DETENT

Combining the advantages of the lever and *detent* escapements and eliminating the disadvantages would be the logical resolution to the quest for the perfect escapement. And this is exactly the approach George Daniels (*b.* 1926), the talented English watchmaker, took when he developed his co-axial escapement (see page 79).

But he was not the first to try to apply the advantages of both into one perfect escapement. Around 1820, Abraham-Louis Breguet's name arises again, this time linked to the lever/detent escapement. There was also an abbot and watchmaker named Bise whose escapement functioned on the same principles as Breguet's did 40 years before Bise tried his. And then there was a Swiss watchmaker by the name of Jules Pellaton who in 1923 patented his version, using a similar principle and mechanism.

John Arnold, eminent English chronometer maker.

The movement of the Hunt & Roskell pocket chronometer. In the parlance of the day, a chronometer was a watch with a detent escapement. In this example the detent escapement revolves with its balance wheel in a tourbillon cage to obviate any errors caused by gravity pulling on the balance, pallet fork and balance spring.

THE CO-AXIAL

George Daniels's escapement functions essentially the same as those who tried to combine the two basic detached escapement types before him. However, he deserves a great deal of credit for understanding the functions of this escapement and optimizing its efficiency and creating the co-axial escapement. He succeeded in minimizing the sliding friction by redesigning and realigning the component parts of this combined escapement, simplifying the design, and improving the transmission of impulse force to the balance—advantages the earlier escapements could not offer.

The advantages of Daniels's co-axial escapement were the elimination of lubrication and the reduction in wear and tear between the component parts and the gear-wheel train. The disadvantages were its higher cost of production and the very high level of precision needed to manufacture the component parts and assemble them. This level of exactness is possible only in a high-volume, mass-production environment where economies of scale can offset the higher costs; and it requires a manufacturing firm with very sophisticated equipment and experience. It is to the Omega Watch Company and its engineers that the credit goes for having the vision, the tenacity, the courage and the manufacturing expertise to adopt the first bold new innovation in escapement design in over 200 years—a design that promises to bring the commercially produced mechanical watch to a new level of perfection in timekeeping accuracy, and has the potential to challenge successfully the quartz watch in the high-value luxury-watch market.

ABOVE
John Kay (1704–99) a British clockmaker who was also a forerunner in the development of weaving looms. In 1733 he devised a machine to clean wool by beating it; the same year he patented the flying shuttle, which increased production in weaving and made a broader cloth than was possible before.

The Industrial Revolution began in Great Britain with the introduction of power-driven machinery in the textile industry during the middle of the eighteenth century. John Kay, a Lancashire clockmaker, invented the flying shuttle, a machine that made all of the movements necessary for weaving textiles and helped lead the way toward mechanization. For the newly developed factory method of manufacture to work it required central power generation and iron. So, with its large coal and iron deposits, along with its numerous established domestic industries, Great Britain was poised to lead the world into a new era of mechanized production.

Factories brought the technological innovations of the era—the machinery and production methods—together with workers into a central location to produce goods far more numerous and with greater efficiency than the hand-powered technology of the cottage industry had achieved. Iron was the raw material needed to make the machinery, and powerful steam engines were needed to drive the machinery. The whole system—from the duration of the workers' day to the timing of production—all revolved around the clock.

The Industrial Revolution spread from Great Britain to Belgium, both being important centers of textiles manufacturing. France began to industrialize a little later, but was waylaid by the French Revolution and Napoleonic Wars. It remained an essentially rural and small-business economy until after World War II. Germany industrialized in the middle of the 1800s. The United States (at the time a colony of Great Britain) was the first outside of Europe to industrialize, following Great Britain's lead.

The success of the Industrial Revolution depended on industry's ability to transport raw materials and finished products over long distances. This need prompted engineers, inventors and builders to develop and expand the use of water transportation in the form of canals and rivers. Improved roadways and turnpikes with harder,

smoother surfaces were developed; and so were rail systems using the newly conceived steam locomotives for power.

The efficient use, scheduling and coordination of these expanding transportation systems depended on reliable timepieces, further enabling the transportation systems to link various manufacturing centers in Western Europe and North America.

ABOVE
A crown-verge watch by John Ellicott, Sr., circa 1705, a creditable and enterprising watchmaker. Ellicott was one of the first to use a center second hand; and to make a very thin watch soon after 1700, measuring only ⅕ inch (½ cm) between plates.

TIMING THE INDUSTRIAL REVOLUTION

Just as the world was coming to need a large number of timepieces to order the coming industrial age, the watch- and clockmakers were themselves poised on the brink of industrialization.

The handcraft era in watch production gave way to the *ébauche* system of mass-producing rough watch movements in small shops in England, France and Germany. Frederick Japy, in the French Jura Mountains, set up a factory to mass produce *ébauches*, putting out 100 rough movements per worker in the same time that it took one worker to produce one watch by the handcraft method. The world was demanding ever-cheaper watches. Middle-class and working-class people found that they needed watches in this brave new industrial society.

RIGHT
The fast-growing American watchmaking arena of the 1870s. The Illustrated London News, *June 19, 1875, shows the vast array of processes involved in the machine manufacturing of watches in America, and features the watch factory building in the center of the plate.*

Mechanized finishing was the next step in industrializing watch production. This involved cleaning up, polishing and assembling the watch movements using a systemized production method. Augmented with some light sanding and polishing machinery, the finishing process was vastly quicker than the entirely hand-finishing method.

In the mid-1800s Vacheron & Constantin of Geneva began experimenting with a standardized layout with the aim of shaping and finishing the components by hand. George Leschot (1800–84) developed the specialized machinery to turn out an assortment of uniform components for watches using the same standard arrangement and part sizes. While the parts were similar enough to require much less individual hand-fitting, they were still not interchangeable. Leschot was on the right track, but his machines were too small and lightweight to maintain the narrow tolerances needed for the miniature parts of watches. Also, the lightness of his machinery meant that the parts had to be machined from soft steel, which then needed tempering to increase its strength and durability; and this tempering process altered the shapes of the parts enough for them to then need further hand-finishing and fitting. Nevertheless, the Leschot machines were successful enough to provide Vacheron & Constantin with more movements than they could use—allowing the company to sell the overproduction as *ébauches* to other watch-finishing firms.

Each step in the process—from *ébauches* through mechanized finishing to mechanized assembly—reduced the price of each successive movement. Watches could be made more cheaply by leaving out some of the finishing, polishing and gilding (or plating) steps. Accuracy of timekeeping suffered in the pursuit of lower manufacturing costs, but it also put watches into the pockets of the low-paid workers who drove industrialization.

British watchmakers resisted mechanization of their watchmaking shops and factories, accepting mechanization only to a degree in their rough movement factories. This kept them out of the fray of cheaper watch production and in the high-quality high-price market, while they bemoaned the cheap, inferior Swiss watches that were taking away their stock and trade in Britain and its colonies. Yet it was the more affordable Swiss product that timed the Industrial Revolution and got workers to the factory on time. Their accuracy on the whole was not that bad, and the lower prices could be afforded on a factory worker's salary—albeit at the cost of a few months of installment payments.

BELOW
Eli Terry, master clockmaker of Plymouth, Connecticut. Terry's production of 4,000 wooden gear clocks between 1807 and 1810 was the first successful mass production in America of a product with interchangeable parts.

THE AMERICAN SYSTEM OF MANUFACTURE

In 1770, at the time of the Industrial Revolution in Britain, the American colonies were British, and industrialization began in America, too. The continent's seemingly unlimited natural resources quickly fueled the iron foundries and steel mills that supported industrialized textile, furniture, hardware and shipbuilding industries, especially in the Great Lakes and Atlantic seaboard areas.

Early efforts to mechanize watch production the late 1830s and 1840s failed for the simple reason that the job was much harder than anyone envisioned, and there were plenty who tried. In early economic development efforts, even whole communities in America would fund the development of a group of would-be watch-manufacturing entrepreneurs, only to see the efforts fail miserably and the resulting factory building sit empty.

In 1850 another watch-manufacturing venture began. Aaron Dennison (1812–93), a watchmaker, joined Edward Howard (1813–1904), also a watchmaker, and Samuel Curtis (1788–1879), an investor (and Howard's father-in-law) to form what was to become known as the Waltham Watch Company. From some of the failed attempts to mechanize watch production these three entrepreneurs put together a team of experienced technicians, mechanics and machinists. It took them 10 years with many false starts and failed attempts, including several bankruptcies, reorganizations and corporate restructurings, to learn that the existing machinery (some of which they had invented themselves) was not adequate for the scale of manufacturing that Waltham needed to succeed. The second generation of American machines that they invented and manufactured were heavily framed, single-purpose structures designed to cut and shape hardened steel. The raw steel was fed into successive machines, each one bringing its cutters to bear until the final product emerged, already hardened and tempered and ready to use.

Instead of trying to make pivots to fit jewel holes, Waltham produced huge quantities of both pivots and hole jewels and then matched pivot to jewel hole, using a statistical approach to a system of matching that

RIGHT
A Dennison, Howard and Davis watch, one of the first 4,000 watches machine mass-produced in the world. Made in Waltham, Massachusetts, by the Boston Watch Company (1853–57) This company was the successor to the American Horologue Company (founded 1850), the first in America to attempt the machine mass-production of watches. This timepiece is one of about 4,000 watches produced before The Boston Watch Company fell into receivership in 1857.

ABOVE
The watch companies known collectively as "Waltham" began in 1850 with Howard, Davis & Dennison of Boston, Massachusetts— the first successful machine manufacturers of watches in the United States. In 1853 the firm became the Boston Watch Company. At this time its assets and innovative machinery were moved to Waltham, Massachusetts. It was again reorganized into the Waltham Watch Company and remained so until its closure in 1957. In this photograph, taken in 1885, the American Waltham Watch Company is represented at a trade exhibition.

negated the slight variances in manufacturing tolerances. Matching hairsprings to balances, Waltham sorted huge quantities of both, by weight and force, and then paired them accordingly, again using statistical averaging instead of adjusting by hand, to get matched pairs that would keep accurate rates of time.

By 1858, the Waltham Watch Company had produced 14,000 watches; and six years later, then known as the American Watch Company, they had produced 118,000. It was the birth of a new industry in the United States. The American Watch Company's success encouraged others to adopt its methods and start producing their own watches: Howard Company in 1857; Elgin Watch Company in 1864; Illinois Watch Company in 1869; Hampden in 1877; and Hamilton Watch Company in 1892; along with about 60 other watch companies that slipped into and out of the picture. In over 75 years these various companies produced 120 million jeweled-lever watches. These numbers seem impressive, but in fact the Swiss were producing far more watches. The difference was the higher quality and lower prices of the American watches.

In 1876 Swiss watch executives visiting the Philadelphia Centennial Exhibition were shocked to learn that the fledgling U.S. watch industry could build a watch at a lower cost per unit than the Swiss imports of the time and with at least a similar degree of accuracy. The components were so accurately made by machine and so finely finished that they came off the assembly line with little need for additional adjustment. Although the Swiss were to remain the predominant manufacturers on the world scene, the executives perceived a real threat from the United States and moved rapidly to install this new American method of manufacturing in their factories.

The lack of a watch-manufacturing infrastructure benefited the Americans when they invented the American method of watch production (with their mass-production techniques, tooling and machinery). But it proved to be their downfall, also. The Swiss method of specialization, coupled with the new American machines and tools that they bought and copied, proved far superior to the Americans' insistence on manufacturing totally in-house. With specialization, a relatively small number of component manufacturers could supply a large number of watch producers, dramatically reducing the cost per unit of producing a Swiss watch.

ABOVE
John C. Dueber began manufacturing watch cases in 1864 at his factory in Newport, Kentucky, across the river from Cincinnati. By 1889 he had purchased the New York Watch Company of Springfield, Massachusetts, and moved it to Canton, Ohio, thus securing his own market for his watch case factory products.

Although the American watch industry was never able to dominate world watch production, it did show the world that uniform mass production of precision watches could be done on an economical and profitable basis. The American system of manufacture left its indelible mark on production, not only in watches, but in all other products made throughout the world from then on. It raised the standard of living and the quality of life in the industrial world by making labor-saving and luxury goods available at a price affordable for nearly everyone.

In watch production, it raised the level of quality that could be reached. An accurate, jeweled-lever watch was now available at a fraction of the cost of the old handcrafted masterpieces of 50 years before. The average American worker could and did afford one. The American watch manufacturer also showed that cheap, dependable, but not-so-accurate watches could be made to sell for the single day's wage of an American laborer.

From the beginning of watchmaking, craft-era watchmakers spent as much of their time repairing watches they had made as they did making them in the first place. Keep in mind that watches were and are portable and are subject to environmental forces such as sudden, unexpected movement, shock, dust, dirt and moisture. Watches run 24 hours a day, seven days a week and 52 weeks a year. This constant use, along with deteriorating lubricants, causes wear on the internal components of the watch. Watches, like clocks, tend to self-destruct as they are wound and used; outside environmental factors along with neglect and abuse simply speed up the process.

It was only the intervention of the watchmaker with his regular, servicing techniques that could prolong the life and use of the watch. The skills needed for repairing were the same as the skills needed for watchmaking. As watches became available to the upper middle class, more makers were manufacturing watches to supply the growing demand; but just as many were spending their time repairing those in use. The watchmaker and the watch repairman were one and the same.

A NEW CRAFTSMAN EMERGES

Watch production had left the individual shop by this time and was concentrated in factories. Individual watchmakers focused on repairing watches that were in use. By the mid-1800s, most adult males in the industrialized West had a watch. Usually, it was just one good watch, because they were still expensive, although attainable for most workers. The individual watch-owner took great pride in his watch. It was personal; it was with him all the time. And once a person had a watch it was more affordable to maintain than to replace when it needed servicing.

ABOVE
An early nineteenth-century watchmaker at work in his handcraft shop in England. This illustration is taken from W. Darton's Book of English Trades *published in 1824.*

A new profession developed within the watch industry, based on the repair and maintenance of watches. Because repairing a watch required all of the skills needed to make it in the first place, these repairmen continued to be called watchmakers. They learned the handcraft methods of their forefathers, apprenticing in the old manner. As more watches were sold, more repairmen were needed and schools of watchmaking developed.

In the United States the first watchmaking school was in the home of a Waltham, Massachusetts, watchmaker, D. D. Palmer (1838–75). The first commercial school of watchmaking was the Parson Institute in La Port, Indiana, founded in 1886. Before this, watchmaking was learned either by apprenticeship or in one of the watch factories.

Graduating from a school of watchmaking gave the new watchmaker the basic skills in a shorter time than under the old apprentice system, but he still needed to hone his skills under the practiced eye of the experienced craftsman. So graduates apprenticed themselves for a couple of years, earning a small wage from the local watchmaker, and then went on to open a shop of their own in another town if they could afford to do so; or they continued to work for others within the watchmaking profession.

The profession quickly grew so that by the start of the twentieth century there was at least one watchmaker in every town and village, usually more; and in larger population areas there was at least one watchmaker for every 5,000 or so people.

One of the unique by-products of the American method of manufacture was the production of a large number of interchangeable spare parts for the repair and maintenance of watches. Up until the 1930s, watch components from Switzerland often required hand-fitting into another movement of the same model. By the 1930s, the Swiss, too, could offer full interchangeability, within each model, of their movements.

This availability of spare parts changed the complexion of the repair industry also. By the 1940s, watch repair schools stopped teaching many of the old handcraft techniques. Making a replacement balance staff (the balance-wheel axle) and the winding stem, and hand-fitting replacement wheel-arbor jewels were about the only craft skills being taught. And even these skills were treated lightly in many of the schools. Most of the students' time was concentrated on replacing and fitting interchangeable replacement parts. The hand–eye coordination

RIGHT
For a brief period in 1851 the Warren Manufacturing Company took the name of the American Horologue Company, but when watches began to leave their factory at Roxbury, the company confidently adopted the name "Boston Watch Co." and factory watch production became a reality in the United States. This watch, marked "Samuel Curtis" after their chief investor, was made by the Boston Watch Company in 1852 or early 1854 at the Roxbury factory, before it moved to Waltham.

ABOVE
An 1896 photograph of workmen in the small press department of the movement shop belonging to the William L. Gilbert Clock Company's factory in Winsted, Connecticut.

was still a much needed skill to be developed, but a lot of the handcraft-era techniques were neglected for lack of demand.

In Switzerland and in England, where the handcraft tradition was much more ingrained in the fabric of their watchmaking industries, the traditional skills continued and continue even today to be passed down to each successive generation of watchmaker. Yet, even in the United States, some watchmakers have continued these traditional skills and methods and can rightly be called watchmakers in the traditional sense. Preserving and restoring clocks, too, has gained a great deal of interest.

And now, at the start of the twenty-first century, there is a renewed interest in learning the ancient skills and methods of both the watchmaker and clockmaker. People cherish the old and ancient timepieces of their ancestors, simply because they were both such close, personal possessions as well as marvels of micromechanical innovation and craftsmanship. In the throwaway society of the Western world, hanging on to quality, stable timepieces from the past gives many of us a tangible link to our past.

THE AMERICAN METHOD IN PERSPECTIVE

To put the American system of manufacturing into perspective, we'll look at the four stages of watch production:

- **1700s–1800s**: The most successful shops during the craft era of the eighteenth and nineteenth centuries produced perhaps upward of 1,000 watches a year.
- **1800–1850s**: Switching to the *ébauche* system of manufacture, relying on outworkers to supply the needed components, and yet still relying on handcraft methods to finish the watches, increased production of the leading Swiss and British shops to around 5,000 units per year.
- **1850s**: Mechanizing production of *ébauches* and component parts, along with mechanized finishing, yet still fitting and adjusting by hand methods, allowed some of the leading makers to produce tens of thousands of watches per year.
- **1860**: The American system allowed the Waltham Watch Company to more than triple the average output of a Swiss watch producer.
- **1900**: Waltham was producing more than 600,000 pieces per year.

Impressive as these numbers appear, the Swiss watch producers still outproduced the Americans, simply because there were far more Swiss producers than there were American; but the Swiss product was of much lesser quality than the American.

LEFT
The Langdon grist mill in Bristol, Connecticut, built prior to 1811. Clockmaker Ephraim Downs purchased the mill in 1825 to provide waterpower for his clockmaking machinery. He reportedly paid for the mill with wooden geared clock movements.

INVENTION'S DEBT TO WATCH- AND CLOCKMAKING

The first mechanical clock probably contained as many new inventions as any device since: using a weight as a driving force; using a set of gears to transmit power; a controller and a reciprocating device to slowly measure out the transmitted energy in equal units; and a friction clutch to set the hands without interrupting the transmission of power.

Those who collect timepieces amass more than timekeeping devices: they possess the source of ideas and inventions upon which all other industries depend. The study of watchmaking and clockmaking is the study of both invention and the developing inventive mind. It is a fact that watchmakers and clockmakers have contributed more to invention than any other source.

For example, the *endless cord* (or endless chain), an important invention for modern mechanisms, was first used by Su Sung to transmit power in the eleventh century. It was later used on chain-driven hoists, and Leonardo da Vinci (1452–1519) shows sketches of endless chains used in early weapons systems. Christiaan Huygens (1629–95), the Dutch physicist, used an endless cord in his pendulum clock to provide a continuous flow of power while the clock was being wound.

The *block-and-tackle* pulley system was used to extend the running time in early clocks. This was another Huygens idea, and could extend the running time of a clock by as much as a month or even a year by compounding the pulley-cord system—a system that has been used since in mechanics to multiply lifting and pulling power.

The *count wheel* in clockwork, which counts the hour strikes, found further applications in mechanical memory devices and in mechanical computers.

Worm gearing, first found in clockwork calendars and astronomical devices, was later used in hoists, speed reducers and critical-control devices, providing a means to advance a gear by a measured fraction of a gear tooth, instead of one tooth at a time, giving fine control to measuring devices and instruments. The worm gear system is used to

Japanese weight-driven temple floor clock of the early nineteenth century. The Japanese kept time temporally, dividing the daylight hours into equal numbers of hours of equal length and the nighttime hours into equal numbers of hours, also. In the summer the daytime hours were much longer than the nighttime hours and conversely in the winter, requiring two foliot oscillators and continual readjustment.

reduce noise in engine-driven gear systems. And in a reverse application it is used as a speed governor in music boxes.

Rack-lever counting (or the idea of sector gears) was used in clocks around 1676 to count out the hours. It was later used in mechanical communication machines and calculators.

The *block chain* was first used in *fusée* watches. We're most familiar with it in its application as a bicycle chain. Invented by a Swiss watchmaker named Gruet in 1664, it is now used in assembly lines, machinery and the heavy-trucking industry to transmit a powerful even flow of power under heavy load conditions.

Variable speed gearing, a clockwork principle, provides for automatically programmed variable speeds in production lines without the need to control the speed of the power source. The system uses gear wheels powering *elliptical, triangular* and *lobed cams*, another clockworks innovation, to vary speeds. Early examples of both the variable-speed gearing and the use of cams can be found in the early astraria and celestial globes.

The *frictional clutch* was used originally to allow the hands of the clock to travel around the dial as the clockworks ran, but then permitted the resetting of the hands without disconnecting them from the continually running clockworks. Although there are many examples of frictional clutches in use today, the most familiar is the manual transmission of the automobile. Depressing the clutch pedal separates the frictional pads on the engine drive shaft from the pads of the wheel drive shaft. Releasing the clutch brings the frictional pads back together, and the engine and drive shaft work together, as if they were one unit.

Late eighteenth-century ship's clock. Its springs, chain and gear wheels are made of hard-wearing steel. The case is of gold-plated brass.

Differential gearing, used in an equation clock by Joseph Williamson (*d.*1725) of London in 1720, showed the difference between solar and mean time. Today, we find the differential gear system most commonly used in the rear end of automobiles, allowing each drive wheel to travel at the same speed when moving in a straight line but then allowing the two wheels to travel at varying speeds when turning a corner—all the while allowing both wheels to maintain traction.

Roller bearings, used in heavy machinery and even in toys, help reduce friction, heat, and wear in running objects. Henry Sully (1680–1728), an English watchmaker practicing in France, used roller bearings in his chronometers in the early 1700s. John Harrison (1693–1776), the English chronometer designer, used them on his early clocks. It was from roller bearings that ball bearings evolved.

Governors to control or govern speed can trace their origin to the stop-and-go action of the clock escapement. Governors have been used to control the speeds of ringing bells, continuously running engines, stem generators, music boxes and star-tracking transits. The steam governor, attributed to James Watt (1736–1819), the Scottish-born engineer and inventor, can be traced to his early apprenticeship as a clockmaker.

The *thermostat* found in toasters, electric blankets, blinking lights, traffic signals, furnaces, air conditioners, fire alarms and many other devices is a contrivance that shuts off energy and/or turns it on at certain temperatures. John Harrison first used the thermostatic principle in clock pendulums in the eighteenth century to compensate for the varying effects of temperature on the length of the pendulum. Varying temperature changes the effective length of the clock pendulum, causing a warm pendulum to expand, lengthen, and swing more slowly; or, conversely, making a cold pendulum contract, shortening its length and causing it to swing faster. Harrison found that by combining brass, which expands twice as much as steel, with steel he could negate or compensate for the effects of temperature change.

Other watchmakers in the eighteenth century used Harrison's thermostatic principle to make compensating balance wheels to stabilize the dimensional change of the balance wheel with changing temperature. Applying the thermostatic principle with electrical circuits permits a make-or-break switch that operates on electrical current and the heat that it generates.

Jeweled bearings are used to reduce friction and to retain oil where metal pivots turn in jeweled holes. Nicholas Facio (1664–1753) invented the jeweled bearing in about 1704 for use in watches; it is now used in all precision instruments in every industry.

Oil sinks are formed around wheel axle pivoting holes to keep oil from spreading away from the pivoting ends of the axles. It was a concept that eluded a number of inventors until Henry Sully (1680–1728) and Pierre LeRoy (1717–85), both watchmakers, worked it out in the mid-1700s.

The *universal joint*, another mechanical invention by clockmakers, is used in automobile drive shafts and other power applications, to transmit power from the engine to the drive wheels through minor angles in the drive shaft. Called a *d'Cardon joint* after its inventor, d'Cardonal, an Italian clockmaker, in 1525, it is still referred to as a Cardon shaft in propellers, and constant velocity (C.V.) shafts in today's automobiles. *Gimbals*, used to stabilize moving objects, are a form of universal joint and have their origin in the clockmaker's shop.

Even the *mainspring* has had numerous applications beyond clocks and watches.

The *modern assembly line*, the *steam engine*, the *steam locomotive*, the *steamship*, the *sewing machine* and the *linotype* can all be attributed to inventors who had their mechanical training and technical education as watchmakers.

Clocks are prototypes of all the greatest mechanical inventions since the beginning of time, not only as timekeepers but as tributes to the ingenuity of man.

The movement of a pocket chronometer, No. 971, with a detached verge escapement designed by LeRoy. The movement and escapement were made by Abraham-Louis Breguet, Paris, 1813.

CHAPTER 6

A MOUNTAIN INDUSTRY EXPLODES

"Time is the most valuable thing a man can spend."

THEOPHRASTUS

ABOVE
Looking south from Lausanne, Switzerland, across Lake Geneva to Mont Blanc where the Swiss, French and Italian borders meet. The tranquil, serene and isolated mountain valleys of the region made an ideal setting for the solitary, intense and intricate work of the watchmaker.

The Jura is a mountain range that forms Switzerland's western boundary with France. It consists of a series of parallel ridges that are separated by narrow valleys. An isolated and tranquil area of mountain forests and open valleys, it is suitable for lumbering and grazing stock cattle and dairy, but for little else. The soil is too thin and the growing season too short to support any farming beyond subsistence crops for the sparse population.

The snow usually comes in October and stays until late May, and in the days before motorized transportation, it would clog the passes, bringing all travel to a halt for the six or seven months of winter. During this time the men and boys of the mountains spent their time mending and repairing their farm equipment, and then spent the rest of their winter making wooden barrels and other wood-crafted items; the women and children spent their time tatting lace—all for export to the outside world. Their travels to Geneva and Berne to sell their winter-made wares brought them into contact with watchmaking, the residents took to this new craft immediately. This was the early eighteenth century. The mechanical arts that they had learned as subsistence farmers gave the mountain residents the background needed in mechanics. The experience of the wives and children in making fine lace easily adapted to the intricacies of watchmaking and made watchmaking a family affair for the long, isolated winters.

Families quickly began specializing, each making specific watch components. Each family would become adept at its own specialty and create simple machines to speed up their work, allowing each to provide a steady supply of components to the *ébauche* assemblers in the mountains. These *ébauches* (rough movements) would then be sold to *cabinotiers* in Geneva for final fitting and finishing. The *cabinotier* was a highly skilled, fiercely independent watchmaker, working in a small space, called a *cabinet*, and turning out finely finished and adjusted

THE CABINOTIER

"The cabinotier is an unusual species of watchmaker, a combination craftsman, scholar, artist and winebibber, a man who may not always have been a genius, but sometimes was and actually behaved like one ..."

Most cabinotiers worked in small "cabinets" or shops lined up under the roof tops, utilizing the bright day-light coming through closely set windows. Most of the cabinotiers worked in the Faubourg area, across the Rhone River in the Old City of Geneva. The atmosphere was unconventional, the village crowded and the hours worked, although long, were not always regular. On a whole, the cabinotier was not only a watchmaker; he had been to the College of Geneva, to art school and was as much cerebral as he was mechanical in his aptitudes and skills. Up in the cabinets, his skilled hands working, his mind was still free to wander: listening to someone reading; conversing with a colleague on any of a number of arcane topics, politics, the arts, or music. Exploring his passion for discovery, novelties and improvements in the art and craft of watchmaking, he made the advancement of his profession his lifelong pursuit; many of them working for a lifetime with scientists advancing the art and science of watchmaking without regard for their own personal glory, but for the inner satisfaction of having contributed to their profession. The cabinotier was a practical joker, fond of strenuous exercise and outdoor activities, good wine, good friendship and good fellowship; and he was fiercely proud of his own profession. He considered himself an artist on an equal par with all other artists, no matter what their medium might be.

(Extracted from "Geneva, the Birthplace of Swiss Watchmaking," an article in *The Swiss Watch* magazine of August, 1958.)

watch movements. The *cabinotier* was the middle link in the Geneva *fabrique*. *Fabrique* translates roughly to "factory," but in the eighteenth century the Geneva *fabrique* was more a system of manufacture in which merchant-manufacturers brought *ébauches* and parts to the *cabinotier* and then collected the finished movements for casing and dialing prior to marketing and distributing the completed watches.

The production of rough watch movements (*ébauches*) flourished in the cottages around the towns of La Chaux-de-Fonds and Le Locle, in the Neuchâtel Mountains of the Jura. Specialized toolmakers emerged in the region, supplying newly designed tools and machines to help mechanize the manufacture of their *ébauches*. The quality of their *ébauches* surpassed the *ébauches* made by low-cost labor in Geneva, and these family-made rough movements, made under the handcraft, cottage system of the Jura, were at an even lower cost per unit than those made in Geneva.

A *fabrique* unique to the Neuchâtel region evolved, where watch components were brought together from the various cottages, assembled, finished, cased and marketed separately from those of Geneva. The Neuchâtel watches were produced far more efficiently, keeping the Geneva watches in the higher-priced end of the watch market.

Just as the Geneva *fabrique* had sought out a cheaper source of watch components from the Neuchatel area, the Neuchatel craftsmen sought sources of cheaper component parts by going to homes and villages farther up and down the Jura mountain range and setting up individuals and families in the specialized crafting of component parts. As the outworkers, as they were called, became more sophisticated and set up their own local *fabriques*, the various makers reached farther and farther up and down the almost 250 miles of the Swiss Jura range for more suppliers of parts to feed the rapidly growing number of *ébauche* makers.

One hundred and fifty years after watchmaking was officially recognized as a professional craft in 1601 in Geneva, specialization was firmly in place in the eighteenth-century Jura. A further forty years later and mechanized *ébauche* manufacture was firmly established there, too. Separate finishing factories were set up in the Jura that finished all of the available Jura-made *ébauches*. The supply of *ébauches* for the Geneva *cabinotiers* dried up and the City of Geneva had to develop its own mechanized *ébauche* manufacture; their practice of using the handcraft method to make *ébauches* just wasn't efficient enough any more.

By 1790 the Jura surpassed the City of Geneva in number of watches produced. In 1814 the City of Geneva became a part of Switzerland and thus a part of the Swiss watchmaking economy. The Swiss watchmaking system of manufacture that evolved in the Jura used a relatively small number of specialized component makers that would supply a large number of *ébauche* makers and watch finishers.

ABOVE
French traveling coach watch of the latter half of the seventeenth century. Too large to fit in a pocket, these watches were suspended from a cord around the neck. The dial has a phase of the moon aperture. This is an early example of a watch with a minute hand, starting to be used around 1675 as the balance spring made watches more accurate.

OPPOSITE
Movement and case profile of the traveling coach watch. Note the flowing, flowery balance cock, typical of French work at the time.

ABOVE

The watch factory of Frederick Japy, Beaucourt, France. Japy had developed a number of specialized tools as early as 1780 for the mass manufacture of Swiss unfinished watch movements. By 1800 this power-driven factory was making rough watch component parts by machine.

Their unique system of specialization allowed the *fabrique* the flexibility of producing huge quantities of a particular style or a limited run of a few specialty models, and allowed them to respond rapidly to changing demands and fashions.

By the mid-nineteenth century the Jura watchmakers produced every kind and quality of watch imaginable, for every market. They copied American-style movements for the American market, English-style movements for the English market, Dutch-style movements for Holland, and French-style movements for France. They produced large movements and small, depending upon the market and the demand; they produced cheap, low-quality, low-priced watches all the way to the finest, most precise and meticulously finished watches for the scientist and the aficionado, who could appreciate the beauty, intricacies and performance of the finest watches made.

THE MACHINE AGE COMES TO THE MOUNTAINS

With Fredrick Japy (1749–1813), the machine age made a big step forward in manufacture that was only hinted at for the watch 100 years earlier. Japy, working in Beaucourt, a French village on the northwestern side of the French Jura, conceived the first viable mass-production watch machinery for making *ébauche* movements.

The machines were small and lightweight, and needed constant skilled adjustment and control. The components were not interchangeable but they were similar and fairly consistent. His factory of some fifty employees turned out blank movements at a rate 100 times faster than the handcraft shops of Geneva. Even though these watches too had to be hand-fitted, finished and adjusted before the wheel train could be assembled and made to run, the era of handcrafting was over and the age of machine manufacturing had begun. By 1800 Japy had a power-driven factory and had adapted the Lépine style of movement for production.

On the heels of Japy, the French government sought to create an *ébauche*-manufacturing colony in the French Jura using Swiss watchmakers. Twenty years after Japy had begun production the Swiss themselves started a mechanized blank-movement factory in Fontemelon, also in the Jura Mountains (but on the Swiss side). By 1845 the Japy and Fontemelon *ébauche* factories were producing a combined half-million blank movements per year.

By 1820 and with the Industrial Revolution in full swing throughout Europe, the Jura Mountains, including Neuchâtel, had achieved world dominance in watch manufacturing—becoming watchmakers to the world, a distinction they would hold for the next 150 years.

Nearing the middle of the nineteenth century, Vacheron & Constantin of Geneva took another step forward by putting watchmaking machinery into their own finishing factory, thus eliminating the need for outside *ébauche* suppliers. This put manufacturing from start to finish under one roof.

Here, too, in the mid-1800s, stem-winding was introduced by the newly formed Patek Philippe company of Geneva, but it would not be until the turn of the twentieth century that a stem-setting mechanism would be in common use. The detached lever escapement that had evolved from the rack lever, through the English lever to the Swiss anchor-lever escapement that we know today, became universally accepted for quality watch movements.

By 1870 the pin set-mechanism to set a watch's hands was introduced; before this they were set using a separate key. Also at this time, another significant event was to impact world watch production, this time arising not from the mainstream watch-

BELOW

Precision pocket watch by Paul Ditisheim. At the end of the nineteenth century, English makers were still producing the world's finest watches. In 1903 Ditisheim of La Chaux-de-Fonds, Switzerland, won first prize (and set the record of 94.9 out of 100) at the British watch performance trials at Kew. By 1907 Swiss watches were taking first prize in the pocket watch category every year. Ditisheim is known not only for breaking the English dominance of the high precision watch, but also for being the greatest watchmaker of his time.

VALLÉE DE JOUX: THE HEART AND SOUL OF SWISS WATCHMAKING

The Joux Valley is an isolated valley 3,000 feet (915 m) up in the Swiss Jura, about 60 miles (96 km) north of Geneva. Even 100 years after the City of Geneva's watchmaking guild was chartered in 1601, the "valley" was a practically forgotten land. A few monks had settled there seeking a tranquil area for meditation and solitude. A few hardy settlers had come seeking pasture and they managed to eke out a living for themselves and their families. The long harsh winters closed the only pass connecting the Joux Valley with the outside world and put a halt to agriculture for a good six months each year. So the farmers of the valley supplemented their livelihoods with their unique blend of ingenuity, motivation and inherent technical aptitude. At first they excelled in woodworking: crafting barrels, furniture and implements from the limitless wood of the surrounding forests, some for personal use but most for export out of the valley. When sources of iron ore were discovered, they built forges and used charcoal as fuel to craft wrought-iron implements and fixtures for export. When suitable sand was found, they made glassworks, again exporting their handcrafted products out of the valley and into the cities and villages beyond.

Their export ventures eventually brought the valley inhabitants into contact with the Geneva jewelry trade. There they learned the art of lapidary, cutting and polishing gemstones for the jewelry makers of Geneva, a far more lucrative activity for the long winter months than any of their previous pursuits, and one that suited their individualism, creative minds and precise, mechanical skills.

Contacts with jewelers introduced these uniquely gifted people to the art of watchmaking. Over time, individuals began journeying to Geneva on Lake Leman or to the Canton of Neuchatel to learn the art, returning to teach it to others. In less than two generations, the art of watchmaking had reached the remotest of homes in the valley. The demand for *ébauches* from the Geneva *fabrique* encouraged the valley craftsmen to apportion the work among their family members, each according to his or her skill. This cottage system of specialization allowed the new *ébauche* makers in the valley to produce their semi-finished watch movements rapidly and economically.

It was a system that suited the valley craftsmen well. They did not have to deal with the intricacies of world

Erotic watch, Swiss-made, circa 1810. The automated lewd scene is concealed behind two spring-activated doors. With the doors closed, the dial is all innocence.

A trade card from the firm of J.J. Badollet of Geneva, circa 1875. Jean Jacques Badollet (1756–1829) was active as a watchmaker from 1770 and founded the firm in 1779. They bought ébauches, *finished them and marketed them, as well as brokering* ébauches *to other finishers. The Badollets were an eminent family of watchmakers and watch firms with an unbroken succession from the sixteenth to the early twentieth centuries.*

commerce and distribution and they could concentrate on what they enjoyed the most—the mechanical artistry and creativity of watchmaking. Their curiosity and talents led them to seek greater and greater challenges in the watchmaking arena, culminating in the making of the most complicated of watch mechanisms. By the turn of the nineteenth century, the valley's craftsmen were known for their specialty timepieces: watches that played music, those that recorded lapsed time, and those that repeated the time to the nearest quarter of an hour on little gongs and bells, all within the watch itself. Today, watch firms throughout Switzerland are still going to the Vallée de Joux for complicated timepieces, just as makers throughout the watch-producing world did two centuries before.

Some makers did leave the valley, setting up watch firms in Geneva and elsewhere; some firms stayed and grew into large *ébauche* and finished-watch manufacturers, such as Jaeger-Le Coultre, Blancpain, Fredric Piguet and Audemars Piguet. Others remained small, very small, still operating out of homes and small shops. Some produced highly complicated mechanisms for larger firms; others crafted individual movements and even cased watches under their own name or for other firms, for the very limited, exclusive market of the wealthy watch connoisseur. Yet all these firms, both large and very small, draw upon the unique atmosphere of the valley and upon the ethnic characteristics and innate skills of its inhabitants. It's no surprise that Vallée de Joux has been long known as the "Valley of Time."

RIGHT
A teaching model of an English detached lever escapement. The T-handle winds a mainspring that powers the large fourth wheel and pointed escape wheel, and gives impulse to the tangential lever and balance wheel. The student can observe the actions of this large model and learn how to repair and adjust the relationships of the components in order to gain the maximum performance from this escapement.

BELOW
A high quality Swiss bar movement, made in La Chaux-de-Fonds, Switzerland, circa 1870. Using a machine-made ébauche*, the movement was finished to a high degree of precision with a high quality escapement, balance and jeweled wheel bearings added.*

producing centers but from an upstart nation with a small watch industry that would never gain world dominance yet it would send shockwaves throughout the industry worldwide.

Both Britain and Switzerland exported to America. Ideas for using machinery to enhance watch manufacture beyond mechanized finishing and standardization of sizes had been discussed or tried at various times in Europe and Britain, but none had displaced the traditional method of manufacture. It was not until 1850, in America, where there was no traditional watchmaking industry, that Aaron Dennison, Edward Howard and David P. Davis set up a company to assemble watches using interchangeable parts manufactured by (as yet to be developed) machinery. There was considerable difficulty in setting up this project but by 1857 they had put out a $20.00 watch that had cost $150,000 to develop. Not surprisingly, the firm was bankrupt. Nevertheless, a whole new system of watchmaking was born.

The Swiss makers quickly took the hint and adapted the American system of watch manufacture to their own system of specialized component manufacture. The outcome was a formidable system of watch production that helped Switzerland maintain its position as the dominant world watch producer for the next 120 years.

COMPLICATING TIME

There is one area in Switzerland that didn't have to adapt to the rapidly changing mechanization of the revolution in watchmaking. This area is the Vallée de Joux, whose unique brand of watchmaker specializes in calendars showing the day of the week, the date of the month, the phase of the moon and the current year; alarm functions to remind you of an appointment; hourly chimes; stopwatch and elapsed-time features; and even a chime to tell you the time in the dark. These additional features are called "complications" and have been added to many fine mechanical timepieces for hundreds of years. They were called complications in the first place because they added a degree of complexity to the ordinary timekeeping function of the timepiece, whether it was a watch or a clock; and they represented a degree of knowledge, skill, fine craftsmanship and artistry beyond the mere making of the timepiece.

The earliest of the complications was the alarm. In fact, it would be more accurate to say that the timekeeper was a complication of the alarm, since it was from the mechanical alarm that the clock evolved. The day, date, month and moon phases were components of the astraria that the early clockworks drove (see page 13). But with the advent of the clock as a domestic timepiece, the date was a complication added to the clock. One of the earliest complications, it required the addition of a wheel that turned daily, tripping a ring of 31 calendar dates, with each successive day's number appearing in a window in the clock dial as the month wore on. Or it required a series of wheels that turned a hand by one revolution every 31 days, with a provision to advance this wheel one day ahead when the month had 30 days.

ABOVE
A pin-pallet lever watch by Roskopf of Geneva, circa 1900. Manufacturers sought faster and cheaper methods of machine mass-producing watches. Roskopf led the way with this timepiece, using stamped nickel plates and gears, punched bearing holes and an over-sized mainspring to overcome the friction caused by the crudely stamped gears and bearing surfaces. The watch is set by pushing the pin on the left side of the case and turning the winding knob.

Day wheels, naming the day of the week, added a similar complication to the clock or watch dial. The year dial was a more complicated feature, using a series of wheels that multiplied the hour wheel times 24 hours in a day by 365 days in the year. The moon-phase dial was a similar complication. It usually had a disk with 59 teeth in its circumference and two pictures of the moon drawn on its face. Visible through an opening in the dial, the disk had one moon on each half to show the moon and its phase as the lunar month progressed through its 29½-day cycle (the disk would turn once every two lunar months, or 59 days). The trick was how to divide a circle into an odd number (59) of evenly divided spaces using evenly spaced gear teeth to make gear trains that would process the revolutions of the hour wheel into month-long or year-long revolutions of a hand.

RIGHT
The Packard Astronomical Perpetual Calendar watch by Patek Philippe, Geneva. When it was made to order for James Ward Packard, complete with a celestial chart of the sky over Packard's home town, it was the most complicated watch in the world and remains one of the most complicated, even today. This watch was the first complicated watch ever to be custom-built to a buyer's specifications. It took five years to make and was delivered to Mr. Packard on April 6, 1927.

ÉBAUCHES

One hundred and fifty years after watchmaking was officially recognized as a professional craft in Geneva (1601), specialization was an established fact: spring, tool and chain makers appeared; and also the makers of *blancs* like the English "blank" movements. This term in French meant a "bare" movement without plating or engraving. These *blancs* were composed of a cage or frame that would support all of the movable components of the watch. Certain assemblies, such as the barrel, the *fusée* and sometimes the wheels and pinions were provided within the frame or cage. A *blanc* with all the gear wheels installed and turning, but without the escapement, was called the *blanc roulant*. The extensive work of completing, installing and finishing was left to the *repasseurs* or finishing watchmaker of the other watchmaking centers or the *cabinotiers* of Geneva, both the elite of skilled watchmakers.

In the watch industry the word "*ébauche*" is a *blanc* movement, or rough movement; it could also be a *blanc roulant*. To the ordinary French person, "*ébauche*" describes any piece of unfinished work, be it a piece of art, a plan or anything else. Since the turn of the twentieth century "*ébauche*" has meant a finely finished movement framework with its gear wheels but without its escapement. They are made available in infinite varieties to assembly shops and factories where the escapement is added; often, the quality and type of escapement determines the overall quality of the finished watch.

In 1927, 12 of Switzerland's independent *ébauche* factories, representing 90 percent of the Swiss *ébauches* made, joined together under a single company (called Ébauches S.A.) to standardize, integrate and streamline the production and distribution of their finely finished *ébauche* movements.

By 1951, Ébauches S.A. were producing 70 percent of the movement frameworks. They had lost about 20 percent of the market to a group of *ébauche* factories organized as "Les Manufactures," who made their own exclusive size and style of *ébauches* and finished the watches completely themselves. Many of these companies could trace their origins to the old City of Geneva, and they appeared to be trying to distance themselves from the mountain industries represented by Ébauches S.A., even within the seemingly united Swiss watch community.

Ébauches S.A. made spare parts available on an easily understood system worldwide, with spare-parts catalogues, identification manuals and interchangeability charts, all developed to make the after-sales servicing of their fine Swiss watches an economical and efficient process for the watch owner.

A typical Swiss movement from the Le Locle area of Switzerland. It is a Lépine-style calibre and was made in the mid-nineteenth century. This style of movement is typical of those being made by the ébauche *factories.*

THE CHRONOMETER TIME TRIALS

The fabric of watchmaking has had one thread that has held true throughout its history: the pursuit of precision timekeeping. From the craft's periods of development, specialization, mechanization and mass production, watchmakers have sought to build an ever more accurate and reliable instrument. Science demanded that they accurately measure and model events in the laboratory; business demanded that they make manufacturing more accurate and systematic; commerce demanded that they ensure more accurate and safer transportation; militaries demanded that they provide for more exact and organized marshaling of troops and armaments.

Early in timekeeping history, the accuracy of a clock was verified by the sundial, then by the transits of heavenly bodies across the sky. Science and watchmaking combined to make possible ever-increasing degrees of accuracy in mechanical timekeeping. The Swiss went about their self-imposed pursuit of accuracy by uniting the laboratory, the university and the watchmaker. To spur the pursuit of precision timekeeping among the various segments of the country's watch manufacturing, the Swiss instituted competitions among the watchmakers. The first competition was held in 1790, in which the judges wore the watches and compared their rates against a sundial at high noon.

By the mid-nineteenth century two new observatories were built for the watchmaking industry, one in Geneva and one in Neuchâtel. They conducted tests for accuracy and granted certificates of performance to those watches that made the grade. In 1870, the time trials, as they became known, were made into annual competitions. The results were published and the winners could use the results of the competition in their international advertising promotions.

The British had their own set of time trials, but they were reserved for marine chronometers and deck watches, with the Royal Navy purchasing the best pieces for their own use.

The Swiss also conducted competitions for inventions, awarding prizes for improvements in the technical area of watch research and development, such as improved escapements, working methods, production processes and compensation technologies for temperature variation and isochronism (see glossary, page 167). Prizes were awarded, the results published and the technologies shared through the unique Swiss patent system.

This system of competition, prizes and advertising propelled the Swiss watch industry ever further into precision timekeeping and made the phrase "running like a Swiss watch" synonymous with the term "precision."

A precision watch by Josiah Emery (1725–97) London, circa late eighteenth century. Emery was an eminent Swiss watchmaker who settled in London and was the first, after Thomas Mudge, to use the detached lever escapement.

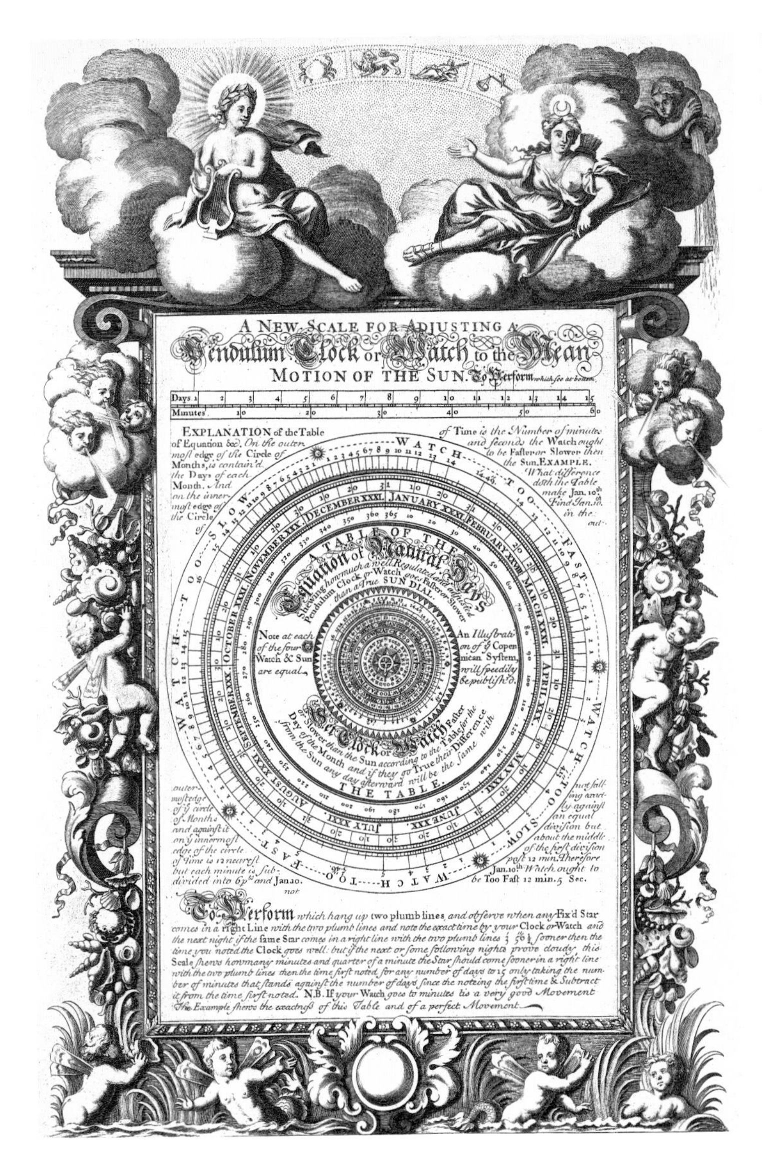

LEFT

Equation of time table, circa 1690. Sun time is not the same as mean time, the uniformly counted time shown by the clock. Sun time runs up to 16 minutes ahead of mean time and up to 14 minutes behind. The two times coincide four times a year. This table was devised so that a person could check the accuracy of a clock against a sun dial and adjust it accordingly, something that had to be done frequently in the early days of rather imprecise watches.

The "equation clock," showing both solar (sun) time and clock (mean) time as a complication, tested the mettle of the clockmaker. The technique called for either a kidney-shaped cam whose outline traced the apparent solar time throughout the year or a complicated set of differential gears which did the same. Variations throughout the year further complicated the task: sun noon arrives about 16 minutes sooner than clock noon at the end of October, and about 14 minutes later at the end of February. Clock time and sun time are equal in mid-April, mid-June, at the end of September and close to Christmas Day. Today, this is an interesting complication for a watch, but in times past it was a necessity when it came to resetting an especially accurate watch with the sundial as the only accurate local time standard, before the advent of the radio and the time signal which was then broadcast by radio on a regular basis.

Equally complex for the clockmaker was the addition of a sidereal time measurement to the clock. Sidereal time measures the rotation of the earth from a "clock star" instead of from the sun. This is more accurate but it ignores daylight, giving a day of 23 hours, 56 minutes and 4.1 seconds in length. Measuring sidereal time is impractical for ordinary daily use, but it is of critical importance in time-determination studies by observatories.

See box on page 122 for more about Complications.

MINIATURIZING THE COMPLICATIONS

It was only a matter of time before a clock's complications were reduced to fit inside the watch. These complicated mechanisms proved too much of a temptation for the watchmaker who wanted to see how small and how intricate a work he could do. Using the skills they had learned while miniaturizing the clock into a pocket watch, the finest watchmakers wanted to show off the ultimate of their skills, adding calendar mechanisms and moon-phase dials along with equations-of-time and sidereal-time complications to their watches. Most watches did not include all of these complications at the same time: some might have the calendar functions with the phases of the moon; others might show sidereal time or the equation of time. But a few watches had nearly all of these complications, making them some of the most complicated in the world.

Some watchmakers added musical boxes within their watch cases; others learned to add small bells or even spiral gong wires that sounded the passing of the hour, and on some the passing of the quarter-hour, just as a chime clock does. The watch that audibly repeated the current time on a series of gongs added one of the most complicated dimensions to watchmaking and was an expression of the watchmaker's highest skill.

In the days before the availability of matches to light a candle away from the hearth (matches became available in Europe in 1855), watches that repeated the time to the nearest quarter-hour were a useful luxury in the long winter nights and in the dark smoke-filled chambers of business and government. Invented by Daniel Quare (1648–1724) in the late seventeenth century, this watch would count the current hour on a gong or bell and then indicate the nearest quarter-hour on a different-sounding bell or gong. Within 20 years the refined mechanism was to repeat the time to the nearest half-quarter-hour, or 7½ minutes. More than 100 years later the minute repeater appeared, allowing the watch to sound the time to the nearest minute on two spiral gong wires inside the circumference to the watch's case.

THE TINIEST OF COMPLICATIONS

Bringing the miniaturization of the complicated watch to its ultimate level of perfection became the goal of the Swiss watchmakers in Vallée de Joux. These watchmakers exhibited a gift for the complicated. They had the mental capacity, the logic, the patience, the determination and the unique handcraft skills necessary to create, design and integrate the finest complications into the pocket watches of the world. And, in the next and final extension of miniaturization, they reduced these complicated mechanisms to fit on the wrist.

Today the Joux Valley is still the source of most of the complications that are added to mechanical watches—complications that represent the finest examples of the watchmaker's art. Much of the making of miniaturized components—the racks, levers, springs and wheels—has been mechanized, but these complex parts need the delicate fitting, finishing and adjusting that only the practiced hand of a watchmaker

BELOW AND LEFT
The Graves Astronomical watch by Patek Philippe, Geneva. In 1928, Henry Graves, Jr. of New York commissioned Patek Philippe to create a watch more complicated than Mr. Packard's. Five years later the watch was delivered. It remained the most complicated watch in the world until the Calibre '89 was unveiled in 1989 to celebrate the Patek Philippe Company's 150th anniversary.

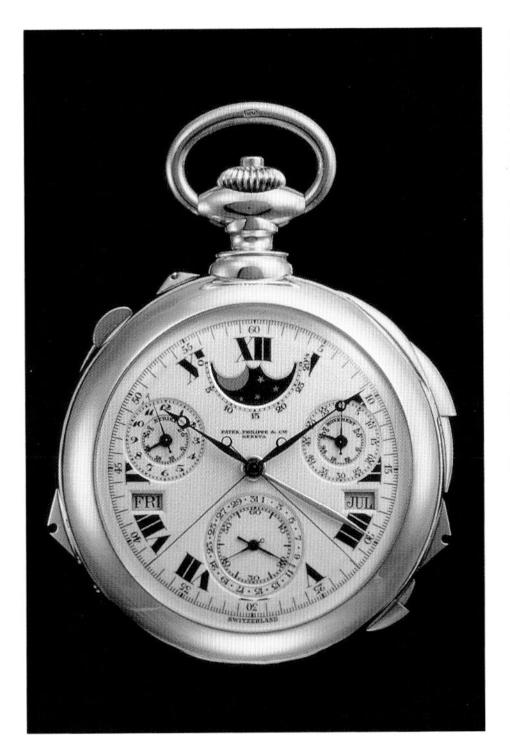

ABOVE
The movement of a gold and leather triple-cased watch by Thomas Mudge, circa 1755. The movement is Mudge's No. 407 with a cylinder escapement and a minute repeating mechanism.

can provide. No machine has yet been designed that can duplicate the gentle, skillful touch needed for these most complex mechanisms. It takes the heart of a fine watchmaker to add soul to these watches—and these complicated mechanical watches certainly do have soul.

WOMEN IN WATCHMAKING

For much of its history watchmaking has been a family affair, concentrated in homes and small shops. Knowledge of the craft and skill development was passed from father to son and...in many cases...to daughter. Women have had a role in watchmaking through much of its history. Unfortunately social and cultural restrictions have limited this role to the more mundane, menial tasks in the craft: the repetitive making, filing and polishing of watch components. In the cottage system women found themselves working side by side with the men making components. In the Industrial Revolution women and children found themselves in the factories turning out watch movement components 12 to 14 hours a day, 7 days a week under poor light conditions. This same sociocultural system prevented women from becoming fully fledged watchmakers and profiting from the work. The old guild system certainly wouldn't have permitted a woman to enter into a formal apprentice program.

It wasn't until the twentieth century that women would be accepted into the watchmaking profession and could seek training and employment as watchmakers and watch repairers. Granted, there were daughters of watchmakers in the eighteenth and nineteenth centuries who successfully worked in their father's shops and went on to practice the craft even after their fathers passed away; but they weren't accorded the recognition and had to practice, if under the jurisdiction of the guild system, with a male watchmaker, often a family member, recognized as the owner and operator of the shop.

Today it is not unusual at all to find women working in the watchmaking and watch repair fields. Their inherent patience, dexterity and generally smaller fingers and hands make them ideal candidates for this profession. To cite one recent example, we'll look at Carole Forestier (b.1968), a watchmaker with the Ulysse Nardin Watch Company of Le Locle, Switzerland.

Forestier is the daughter of a Paris watchmaker. She trained as a watchmaker-repairer at the technical college in La Chaux-de-Fonds (La Chaux-de-Fonds Technicum) and then trained there also as a technician designer in watchmaking. She worked as a designer at a watch engineering consultancy and then as a designer in charge of the technical and prototype manufacturing department of Renaud & Papi S.A. of Le Locle. Since 1997 she has been in charge of the technical development department at Ulysse Nardin S.A.

In 1999 Forestier created a watch that received the first prize in the "Innovation of Mechanical Timepieces," an international invitational competition organized to commemorate Abraham-Louis Breguet's 250th anniversary. Her watch was a "center tourbillon carousel," meaning that the watch's balance wheel and escapement are carried in a cage that rotates once an hour like the traditional tourbillon invented by Breguet himself, but the watch's gear train is also carried in this same cage.

The tourbillon, itself, is a special watch escapement mounted on a platform that revolves, usually once a minute, evening out some of the positional errors inherent in the ordinary watch escapement. An ordinary watch will run at a different rate (or speed) when the watch is lying on a table than when it is on the wrist.

It will also run at a different rate if the winding knob (or crown) is pointed down to the ground rather than in a position horizontal to the ground. Modern hairsprings and balances have reduced these differences in what are called positional errors a great deal, but the rotating tourbillon reduces these errors even further by averaging out the errors each minute that it rotates. Carole Forestier's "carousel" arrangement carrying the gear train along with the balance wheel and escapement is a technical innovation that demonstrates the outstanding skills of this highly talented watchmaker.

THE COMPLICATED WATCH

Each invention, each innovation and each technological advancement in timekeeping represents a masterpiece—but the finest, most complex masterpieces of chronometry are the "complicated" watches, representing the ultimate blend of science, technology and mechanics with the craftsmanship of the watchmaker.

Complicated watches are timepieces that include any number of "complications" or mechanisms that add additional time-related functions to the watch. They can be as simple as a calendar date ring that is tripped by the hour wheel of the watch, or as complex as a perpetual calendar that mechanically calculates the varying lengths of months and leap years every four years without the need for adjustment to the date for decades. A timepiece with a simple calendar function is considered a watch with a complication; if it shows the day, date and year with a moon-phase dial, it has four complications. If it also has a means to readjust the calendar to automatically compensate for months shorter than 31 days, it has another complication.

Complications can be divided into five areas of function: timekeeping, astronomical, chronographic, chime and operational, all added to the basic timekeeping movement.

TIMEKEEPING FUNCTIONS

Examples of timekeeping functions include dials and hands that show solar time, sidereal time or the time zone of a neighboring country or state. They can even include special time-regulation devices to regulate the timekeeping rate in the different positions in which a watch is carried; or special escapements such as the *detent* (see page 82).

ASTRONOMICAL FUNCTIONS

The astronomical complications include the simple day, date and month indicator mechanisms and also a perpetual calendar showing the year within its four-year cycle. A complication showing solar time can be added along with one showing the phases of the moon. Also possible are complications displaying the times of equinoxes, solstices or seasons of the year, along with the signs of the zodiac and the movable dates for Easter.

Dials showing the times of sunrise and sunset for a particular location throughout the year and even a star chart for specific latitudes showing the positions of visible stars throughout the year are all complications that have been designed, calculated and made for some watches.

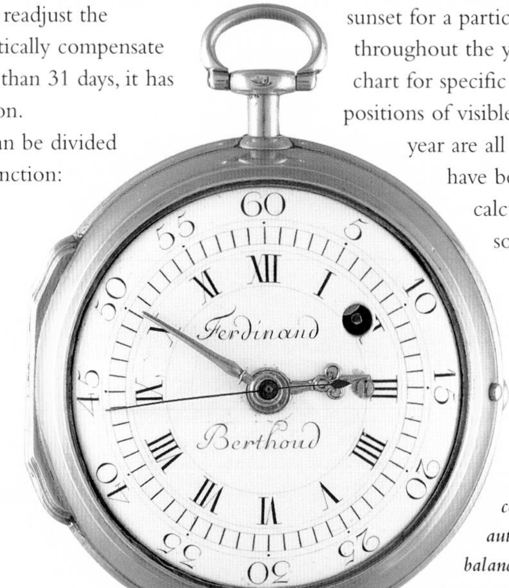

A gold-cased crown-verge chronograph with a grid-iron temperature compensation system to automatically adjust the balance to changes in external temperature. Made by Ferdinand Berthoud, Paris, 1763.

CHRONOGRAPHIC FUNCTIONS

Chronographs, also called stopwatches, are timepieces that measure and record time intervals within seconds and even fractions of a second. They use an additional second hand, which starts and stops with the push of a button.

To measure an event, for example, the button is pushed at the beginning of the event and then pushed again at its end. The position of the stopped second hand on the dial gives the reading of the elapsed time in seconds. If the event lasts more than 60 seconds, a *minute-recorder* dial records the number of elapsed minutes, and an *hour-recorder* dial can record the number of elapsed hours, if the event lasts that long. Each of these recorders is an additional complication, requiring a separate mechanism, yet integrated into one.

CHIME FUNCTIONS

Several interesting complications use chimes. The Grande Sonnerie clock or watch strikes the hour on gongs or bells before chiming each quarter-hour. For example, when the watch is at 3:30, this complication would strike three hour tones and then two quarter-hour tones; at 3:45 it would strike three hour tones and then three quarter-hour tones on the gongs or bells. The Petite Sonnerie strikes each passing quarter-hour, but does not ring the hours except at the top of the hour.

Repeaters are very complicated mechanisms that sound the current time on gongs or bells with the push of a button or the slide of a lever. *Quarter-hour repeaters* strike a deep note on the hour and two shrill notes on each quarter-hour. *Five-minute repeaters* strike the current hour and one shrill note for each five minutes of the current time. The *minute repeater* strikes the current hour, the quarter and the nearest minute, using a combination of strikes and tones.

OPERATIONAL FUNCTIONS

Operational complications can include an up-and-down indicator to show how far the mainspring is wound and levers to turn the Grande Sonnerie and Petite Sonnerie on and off, along with such non-time-related complications as a thermometer and barometer gauge on the watch.

To design and make any one of these complications is in itself an accomplishment of great skill and knowledge. To put several together within the confines of a pocket watch's case is truly a feat of genius.

Today's watch industry is experiencing a rebirth of mechanical watches with complications beyond the easily made calendar and moon-phase complications. Current advances in technology, with a number of very gifted watchmakers willing to devote the long, intense hours required to make and integrate the various complications, are once again making complicated watches for the discerning client.

A gold-cased crown-verge watch by Julien LeRoy of Paris. Made in 1740, the watch will repeat the time on a gong to the nearest quarter-hour with the push of a button.

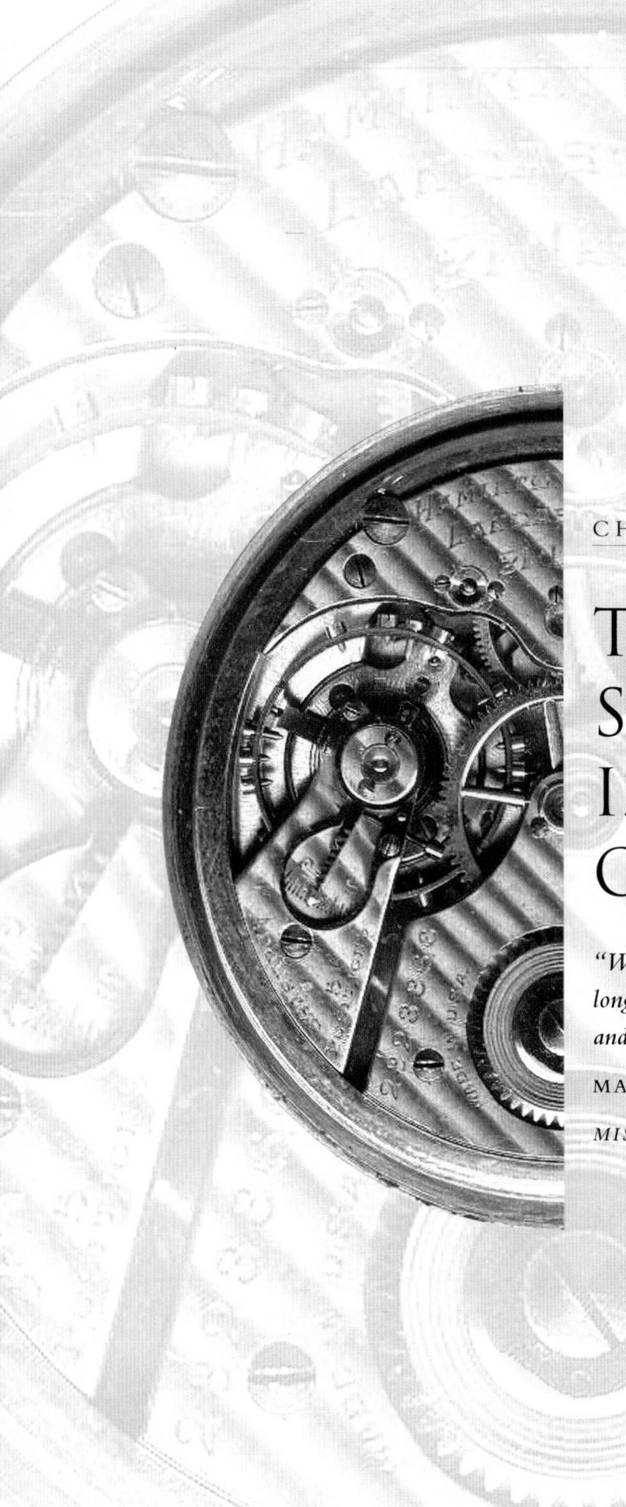

CHAPTER 7

THE STANDARDIZATION OF TIME

"When I'm playful, I use the meridians of longitude and parallels of latitude for a seine, and drag the Atlantic Ocean for whales."

MARK TWAIN, *LIFE ON THE MISSISSIPPI*

ABOVE
"The Wreck of the East India Company's Dutton, *26 January 1796," by Thomas Lung (1759–1837). The painting dramatically illustrates the hazards of sailing by guess and "dead reckoning." It was numerous shipping disasters like this that led to a universal search for a practical method to locate longitude at sea.*

No one knows for sure what caused that Great Age of Discovery that burst upon the world in the mid-1400s but regardless of the cause, Europeans learned more about the world in the next 300 years than had been known during the previous millennia. The spices and silks of the Orient were undoubtedly an impetus. With the Turks blocking the well-known trade routes to the east, the various seafaring nations of Europe sponsored explorations to the west, hoping to find an alternate direct ocean route. The results were new lands to colonize and claim; and untold riches and peoples to exploit.

The coffers of the kingdoms of Europe were bursting at their seams; shipbuilding was proceeding at a frantic pace. The cost: thousands of seamen, hundreds of ships and billions of dollars in cargo (in today's dollars) were lost to the poorly charted expanses of the world's seas and oceans.

Storms, fierce winds and unexpected crosscurrents pushed ships off course and, once a ship was off, the helmsman had no idea how to get it back on course. There were no landmarks in the vastness of the ocean. Those islands and coastlines that did pop up did so literally, because the crew had no idea of their locations—or their own location, for that matter. Many hundreds of ships ended up on rocks and shores that either appeared before they were expected or were never anticipated in the first place.

THE CHALLENGE OF THE AGE

It was the great scientific problem of the age: how to find one's position at sea. For more than two hundred years scientists, astronomers, seamen and governments themselves had been trying to solve the problem. From Galileo to Newton the scientific establishment had sought to map the heavens in order to locate points on land and sea. They divided the earth into parallel lines of latitude running from east to west around the world, measuring the distance north and south from the equator. Finding one's latitude was quite easy: a matter of finding the height or altitude of the Pole Star (Polaris, the North Star) above the horizon. When measured in degrees, the altitude of the Pole Star nearly equals the local latitude on the ground.

Since Roman times an instrument called a quadrant had been used for this purpose. Variations of the quadrant—the cross-staff, back-staff and mariner's astrolabe—were developed in the sixteenth century to make it easier to measure latitude aboard a rolling ship. South of the equator the Pole Star is not visible, so techniques were developed to measure the altitude of the noontime sun and convert this measurement to the local latitude. These instruments and techniques were known and used throughout the Age of Discovery. In fact, it was common practice for mariners to traverse the oceans by immediately sailing north or south to the latitude of their destination and then following that latitude east or west.

The vexing problem plaguing sailors and astronomers alike was the longitude. To pinpoint a specific location on the earth one needs both coordinates: latitude and longitude. Astronomers and map-makers had divided the circumference of the globe into 360 equally spaced lines of longitude running from the north to the south representing the

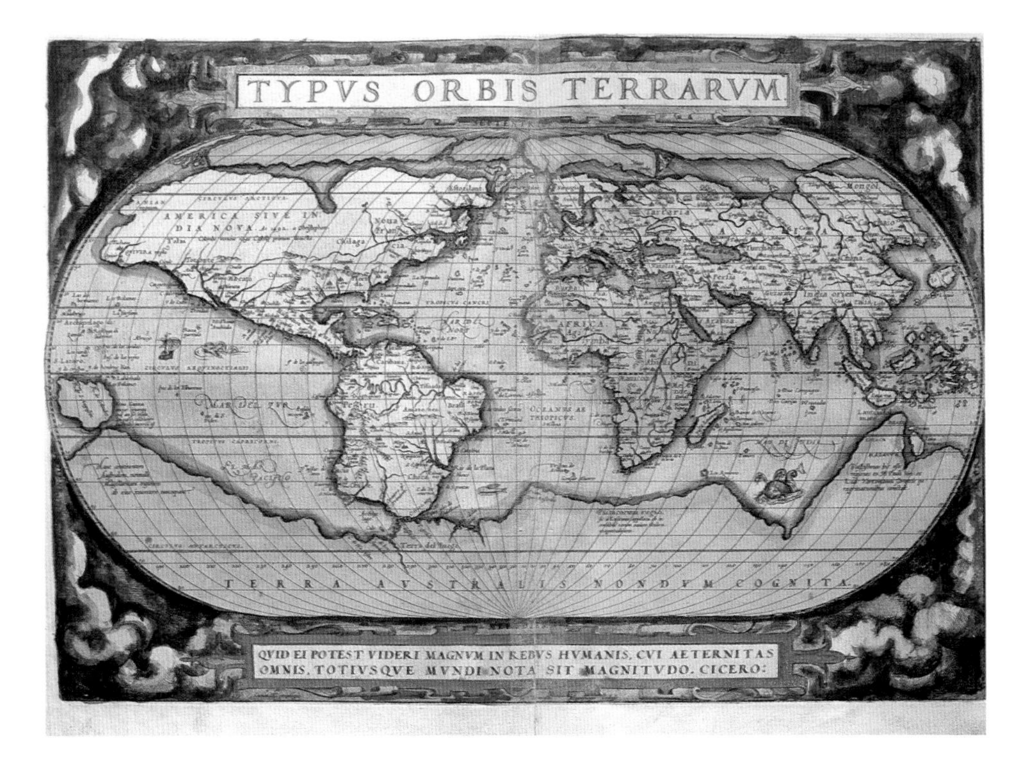

BELOW
World maps of the era were mere approximations. Without a knowledge of longitude, a navigator could only hazard a guess to his location on the globe's surface...and that guess was usually miles off the target.

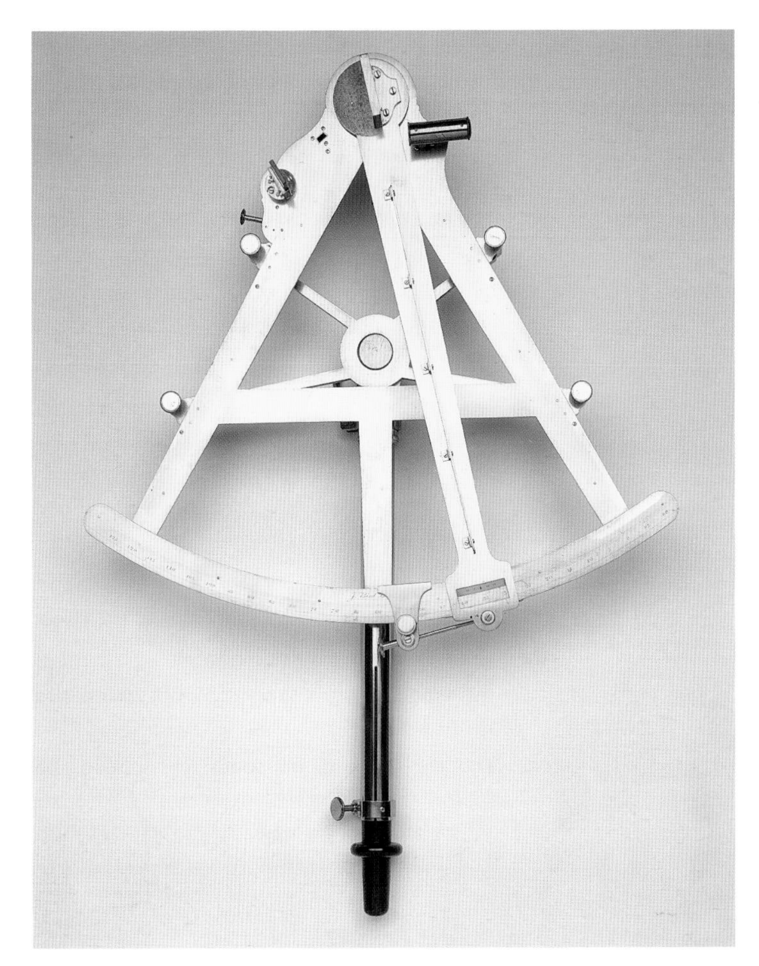

LEFT
The first sextant made by John Bird, London, and ordered by the Astronomer Royal, James Bradley, in 1757. The sextant is the primary celestial navigation instrument, so-named because it contains an adjustable metal frame representing one-sixth of a circle. The sextant is used to measure the altitude of celestial bodies above the visual horizon. By consulting a chart of angular observations taken at numerous locations, the navigator can locate the latitude line he is on (one of the two coordinates needed to pinpoint his position).

360-degree circumference of the earth. (Every 15th line of longitude was called a principal meridian of longitude, so that the earth could be divided into 24 principal meridians.) These 360 lines of longitude could be further divided, like latitude, into 60 one-minute degrees and 60 one-second degrees of arc.

Longitude was measured from a prime meridian, usually the meridian of longitude that ran from north to south through the port from which the ship would sail; later it would be the one that ran through a principal capital such as London or Paris. In fact each of the seafaring nations had its own principal meridian from which they counted east and west distances.

Since the earth makes one complete revolution around the sun in 24 hours, each of the 24 principal meridians represents one hour of distance from the adjacent one. New York City happens to fall very near the 5th principal meridian (the 75th meridian) counted from the

(355)

Anno Duodecimo

Annæ Reginæ.

An Act for Providing a Publick Reward for ſuch Perſon or Perſons as ſhall Diſcover the Longitude at Sea.

Whereas it is well known by all that are acquainted with the Art of Navigation, That nothing is ſo much wanted and deſired at Sea, as the Diſcovery of the Longitude, for the Safety and Quickneſs of Voyages, the Preſervation of Ships and the Lives of Men: And whereas in the Judgment of Able Mathematicians and Navigators, ſeveral Methods have already been Diſcovered, true in Theory, though very Difficult in Practice, ſome of which (there is reaſon to expect) may be capable of Improvement, ſome already Diſcovered may be propoſed to the Publick, and others may be Invented hereafter: And whereas ſuch a Diſcovery would be of particular Advantage to the Trade of Great Britain, and very much for the Honour of this Kingdom; But beſides the great Difficulty of the thing it ſelf, partly for the want of ſome Publick Reward to be Settled as an Encouragement for ſo Uſeful and Beneficial a Work, and partly for want of Money for Trials and Experiments neceſſary thereunto, no ſuch Inventions or Propoſals, hitherto made, have been brought to Perfection; Be it therefore Enacted by the Queens moſt Excellent Majeſty, by and with the Advice and Conſent of the Lords Spiritual and Temporal, and Commons in Parliament Aſſembled, and by Authority of the ſame, That the Lord High Admiral of Great Britain, or the Firſt Commiſſioner of the Admiralty, the Speaker of the Honourable Houſe of Commons, the Firſt Commiſſioner of the Navy, the Firſt Commiſſioner of Trade, the Admirals of the Red, White, and Blue Squadrons, the Maſter of the Trinity-Houſe, the Preſident of the

4 Uuuu 2 Royal

ABOVE

The first page of the Longitude Act of 1714, known as the "Act of Queen Anne: An act for providing a public reward for such person or persons as shall discover the longitude at sea." Although not the first reward offered, the act authorized Parliament to grant a king's ransom of £20,000.

prime meridian that runs through Greenwich in London. If the sun is directly overhead in New York City, it was directly overhead in London five hours earlier; when it is noontime in New York it is 5 P.M. in London.

All you had to do to find out where you were in the world was to subtract your local time (wait until the sun was directly overhead) from the time at some prime meridian—London or Paris, for example. The number of hours and fractions of an hour in difference would tell you precisely how many meridian lines and fractions of a meridian you were from the prime meridian. If it was noontime where you were and it was 4:15 P.M. in London, you were 4¼ principal meridian lines west of London. Since there are 15 lines of longitude in each meridian, you were 63.75 degrees or about on the 64th line of longitude. From the map you could see that this longitudinal line runs through Nova Scotia, Bermuda and Puerto Rico. After finding the angle of the sun or the angle of the Pole Star at your location (and referring to an almanac of sun and star positions and their angles above the horizon at the different times of year), you could tell which latitude line you were on. Where these longitude and latitude lines intersect was your exact position on the earth, and how far you were away from where you wanted to be—or perhaps where you *didn't* want to be.

Simple! Except there was no clock or watch in the world accurate enough to keep correct time aboard a pitching, rolling, storm-tossed ship. The earth rotates 1 degree of longitude every 4 minutes. If a ship was sailing to the West Indies and the captain's watch was accurate to within 4 minutes a day, he would miss the island by a good 60 miles—and watches made during much of the sixteenth century were not only too expensive for a sea captain to own but they easily lost a quarter-hour per day on dry land. Accuracy on water was much worse.

Such misses were common. Mariners traveled by inaccurate charts, using dead-reckoning, which wasn't much more than experienced guesswork, and following coastlines and latitude parallels as much as possible, hoping to spot land masses they recognized—and hoping not to be pushed off the latitude line by some cross-current, drift or storm.

Each successful voyage brought back a little more information to add to the charts of land masses, islands, reefs and shoals. But, without knowledge of longitude, the maps were distorted, especially in the earth's westward reaches, and of only general value. Most mariners got lost on their journeys across the empty oceans, placing them in danger of running out of supplies, becoming ill from lack of fresh water and fresh fruits, or sailing into unnoticed and unforeseen hazards.

A KING'S RANSOM FOR A CURE

Each of the seafaring nations offered a substantial reward for finding longitude: Philip III of Spain in 1598, shortly after the Spanish Armada ran aground off the Irish coast; the Dutch Republic in the early 1600s; France and others followed suit. But it wasn't until 2,000 men were lost, when a squadron of British men-of-war ships ran aground on a foggy night in 1707, that the Royal Navy and Merchant Marine pleaded to Parliament for help. The help came in the formation of the Board of Longitude in 1714 to oversee the granting of a king's ransom of £20,000 (around 12 million U.S. dollars today) to the person or persons who could solve the most vexing scientific and technological problem of the age. The prize was open to all nationalities, and any method with merit to solve the longitude problem would be reviewed. The Board of Longitude was to continue for more than 100 years and distribute over £100,000 in grants to

BELOW
A seventeenth-century painting of the Royal Observatory, Greenwich, from Croom's Hill, soon after its completion, by an unknown artist. Shown within the observatory grounds is an 80-foot (24-m) mast that was used for supporting and adjusting the height of a 60-foot (18-m) refracting telescope.

ABOVE
The Royal Observatory's "Camera Stellata" (Star Chamber). Now called the Octagon Room, it was used for observing comets, occulations of stars by the moon, and eclipses of the sun, the moon and Jupiter's satellites. The telescopes and other instruments were moved from window to window as needed and the observations were timed by the two astronomical regulators made by Thomas Tompion of London, 1676.

support research and development into the longitude problem, but it never awarded the Grand Prize.

With a purse this large, the challenge was fraught with shams and preposterous schemes, but three areas of study emerged as having valid possibilities: terrestrial, celestial and mechanical. The first involved a study of the earth's magnetic fields and then the measurement of the varying strengths of the magnetic fields with a compass at locations around the world. The theory seemed reasonable until investigators discovered that the earth's magnetic fields shifted from place to place unpredictably, and so the project was abandoned.

The celestial method was the one preferred by the established scientific community. It relied on knowing that a specific celestial event would occur at a specific reference longitude, and then comparing its occurrence to that of the local longitude. Knowing the earth rotates 1 degree every 4 minutes, the difference in longitude could then be calculated once the time difference was known between the two places. The celestial method was to use the moon as its hour hand and various reference stars as its dial. Several methods of using the heavens as a clock were tried: lunar eclipses, transits of various stars across a meridian of longitude, the regular and frequent eclipses of Jupiter's moons. But the lunar-distance method proved the most worthy of the celestial methods: measuring the distance of the moon from the stars as it revolved around the earth.

The mechanical method of finding longitude was the simplest method to use, in theory. It was a straightforward comparison of the local time on the ship's clock (set each day when the sun was at its highest point—noontime) and the time on an onboard clock that was set to a reference longitude's time. This would give the ship's location easily at any point in the voyage.

Neither the lunar-distance nor the clock method (known also as the celestial and mechanical methods) was viable in the 1500s and it took another 200 years of intense investigation and understanding of the earth's relationship to the heavens before theory could be turned into practice. The Board of Longitude, created in 1714, had been in existence for just a few years when it became convinced that the lunar-distance (or celestial method) was going to be the one used. The observations and calculations needed to determine longitude this way required far more skill, knowledge and education than the average sailor had or could get. Only scientific voyages could afford to have a professional astronomer on board to do the extremely complicated calculations and take the very difficult angular measurements. The scientific community continued the study of the heavens, mapping the stars in relation to the earth and charting events and their angular relationships to the earth's surface, hoping to systematize and simplify the method.

Several of the finest clockmakers pursued the mechanical timekeeper method instead of the celestial, only to throw up their hands in utter defeat. The pendulum clock, the only reliable timekeeper of the day, would not run on board ship, even in calm waters, no matter how it was balanced, suspended or cushioned against the rolling actions of the waves. The pendulum simply stopped or was erratically thrown about inside its case.

ONE MAN'S TALENTS ...

Yet one man dedicated his life to finding and perfecting a seagoing timekeeper, even in the face of overwhelming odds, criticism and ridicule from the scientific community and the public. After all, to meet the requirements of the Board of Longitude, a clock, whether it be mechanical or celestial in origin, had to be accurate to within two seconds per day and had to be capable of maintaining that rate under the rigorous conditions of an ocean voyage. This would mean that the clock had to be 99.9976857 percent accurate, considering there are 86,000 seconds in one day—an impossible task! So they all thought.

His story is a story of epic proportions. It is a human saga filled with heroism, deceit, dishonesty, brilliance and a single-minded determination. John Harrison (see Chapter 2) had reason to believe that he could build a sea clock that could do more than any other clock

BELOW
John Harrison—the man who solved the greatest scientific problem of his time—pictured with his H-4 chronometer, which could keep such accurate time at sea that a ship's longitude could be accurately determined (yielding the second coordinate needed to locate a specific location on the face of the globe).

THE HARRISON TIMEKEEPERS

H-1, built between 1730 and 1736, took six years to complete.

Sea clocks (**H-2**), built between 1737 and 1739, took two years to complete.

H-3, built between 1741 and 1760, took 19 years to complete.

H-4, built between 1755 and 1760, took five years to complete.

Sea watches (**H-5**), built between 1760 and 1770, took five years to complete.

had been able to do on land. He had already built two clocks that kept time to within one second a month, an accuracy far beyond anything anyone else had even imagined.

Harrison was a carpenter by trade, but, having been blessed with such a remarkable mechanical aptitude, he was drawn to clockmaking. He essentially taught himself the craft, using the material he was most familiar with—wood for the plates, wood for the wheels and wood for the arbors of the wheels. He had a natural curiosity and the mind to follow it. His vision of clockmaking was not clouded or impaired by the prevailing practices learned by the clockmakers of his time. He intuitively knew that friction was the enemy of the precise timekeeper and that lubrication was only a temporary cure for friction because it periodically dried up and further impeded the mechanism.

He experimented with various materials for arbor-bearing holes and came up with lignum vitae, a naturally oily, high-density wood from the tropics which eliminated the need for oiling the wheel arbors. He created nearly friction-free meshings of wheel teeth to wheel teeth by employing "trundles" or rolling-gear teeth that engaged with solid teeth. He understood the reactions of materials to temperature and moisture. As a carpenter he instinctively understood wood: The gear wheels of his clockworks were made of laminations of wood that utilized the inherent strength of the wood's natural grain.

Harrison and his brother James devised a mechanism to keep their clocks running while they were being wound. They invented the gridiron pendulum—a pendulum constructed of brass and steel to take advantage of their opposing expansion-contraction properties to make the pendulum resistant to temperature fluxes and give it a constant, unchanging length. They also invented a friction-free escapement that would count truly equal seconds and did not require oil.

...HIS CLOCKS

ABOVE
Sir Edmund Halley, Astronomer Royal of England, pictured here at age 80. Halley was the first to map the southern sky and to correctly calculate the return of the comet that bears his name. By virtue of his position, he was an ex-officio member of the Board of Longitude. It was through his recommendation that John Harrison came into contact with George Graham and subsequently received Graham's financial support for the first Harrison sea-going clock. Although Halley was committed to an astronomical solution to the longitude problem, it was he who gave continued support and encouragement to Harrison and his mechanical solution.

It is no wonder that John Harrison felt that he could make a portable clock for use at sea when he approached the Board of Longitude with his plan. Harrison's story is filled with the good fortune of being supported by George Graham (1673–1751), the pre-eminent clockmaker of his day; and by Dr. Edmund Halley (1656–1742), of comet fame. But disappointment dogged him from the moment the Board of Longitude rejected his first clock's ocean-going test to the West Indies—a clock that it took Harrison and his brother six years to create. They instead used an abbreviated voyage from London to Lisbon. On that trip the clock, called H-1, showed the captain that he was just 1.5 degrees west of where he thought he was, proving to all concerned that H-1 was a practical solution.

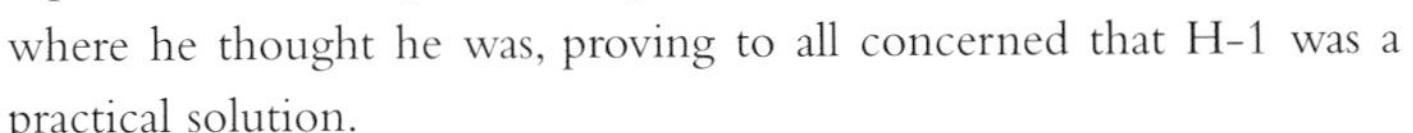

However, something happened during the short voyage that Harrison felt he needed to overcome. We don't know what the problem was, but suspect that it was the clock's size. Measuring about 4 feet (1.2 m) square and 4 feet (1.2 m) high in its gimbaled box, H-1 took up far too much space in the ship's already cramped quarters. Harrison and his brother created a second timekeeper called H-2, which took two years to make, this time using brass and steel for the wheels, trundles and arbors, instead of wood. But during the course of making H-2, Harrison came up with an even better idea, which he pursued, and H-2 never went to sea.

The better idea was H-3, a very complicated instrument having 753 parts and using two large balance wheels controlled by a single, spiral hairspring. It was so complicated and needed so much continuous fine adjustment to keep it going that Harrison put it aside and began working on H-4, a seagoing watch, rather than the previous three seagoing clocks. Harrison discovered during his work with H-3 that a faster-moving circular balance wheel was more stable than either the counterbalanced pendulums (or bar-shaped balances) of H-1 and H-2 or the slower-moving circular balances of H-3. Both H-3 and H-4 were ready for sea trial in 1760, and although H-3 took Harrison 19 years to make, it never went to sea, either.

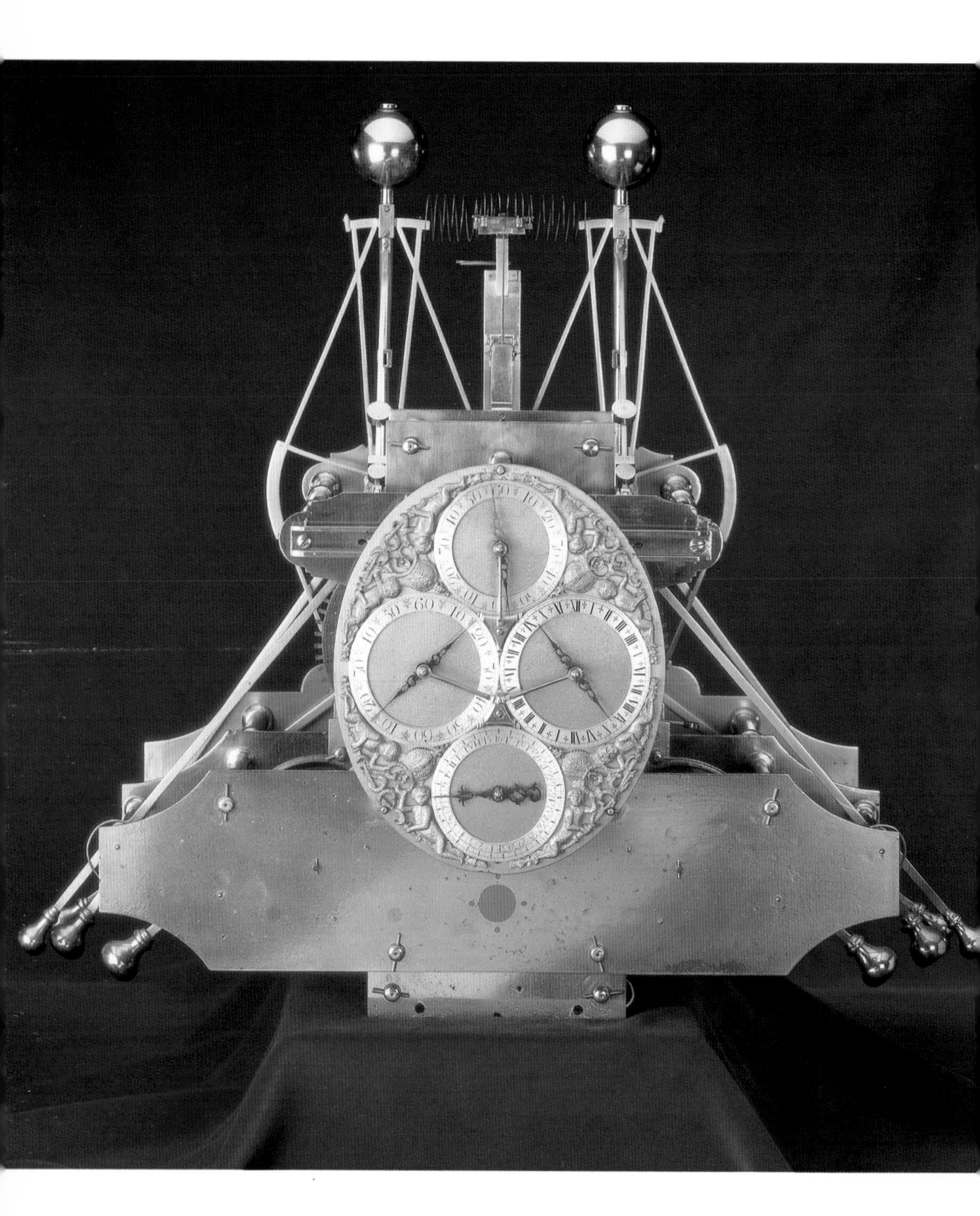

ABOVE
John Harrison's H-1 marine timekeeper, constructed between 1730 and 1735. Its moving parts are controlled by springs, negating the effects of motion on the clock—a necessity in turbulent seas to counteract the rolling motion of the ship.

H-4 was truly a remarkable instrument, not only in its accuracy, but in its artistry, as well. It was small, measuring 5 inches in diameter and weighing only 2½ pounds (0.9 kg), compared with H-1 which measured 2 feet (61 cm) square by 1½ feet (45 cm) high without its case and gimbals; and weighing over 70 pounds (32 kg). H-2 was similar to H-1 but weighed in at nearly 103 pounds (47 kg); and H-3, which weighed something under 70 pounds (32 kg), stood just less than 2 feet (61 cm) high and measured 15⅓ inches (38.8 cm) wide by only 7 inches (17.7 cm) deep.

BELOW
The timekeeper that won the Longitude Prize, known as John Harrison's H-4. To test the watch Harrison's son, William, sailed with H-4 aboard the Deptford *bound for Jamaica on November 18, 1761. After 10 days at sea, they were in the latitude of Moderia, but unsure whether the island lay east or west. The captain estimated they were well to the east. William stated that according to H-4 they were 100 miles nearer to the island and if they stayed their course, they would see Moderia the following morning. The captain abided by the watch and much to their relief the island appeared according to William Harrison's prediction.*

... HIS OBSESSION

H-1 took John Harrison and his brother six years to build; H2 took them two years; H-3 took an amazing 19 years to make and perfect; and H-4 took five, although work on some clocks was done concurrently with others. All of the timekeepers were seaworthy and capable of keeping quite accurate time, but Harrison was obsessed with creating the finest and most practical timekeeper. There were no technologies to support the visions that John Harrison had in creating these instruments. They were all cutting-edge inventions and employed numerous never-before-imagined innovations. To create the technologies to support them, Harrison had to employ his own superbly empirical mind: trial and error, controlled experimentation, observation; creating the methods needed to find the results that he wanted. It took time, a lot of time, and this explains the 30 years it took for John Harrison to make the four timepieces. With each subsequent timepiece, he built up his knowledge and created a more precise, dependable and practical instrument.

There was certainly no question in Harrison's mind and probably no question in the Board's collective mind that any of these instruments could have performed with the precision necessary to win the Longitude Fortune and solve the longitude problem. But personalities and prejudices got in the way. The Board,

ABOVE
Captain James Cook (1728–79). While making three long voyages of exploration and experimentation before his violent death in Hawaii, Cook took the more promising methods of finding longitude to sea with him to test their accuracy and predictability. His praise for Larcum Kendall's copy of Harrison's chronometer (H-4) known as K-1 helped contribute to the rapid growth and popularity of sea-going timekeepers.

made up of establishment scientists for the most part, wanted to see the lunar–distance method succeed, so they put impediments in Harrison's way. They added requirements that weren't in the original set of prerequisites and made the sea trials of his clock very difficult to schedule and carry out. Yet his H-4 watch passed with flying colors. Remember that H-2 and H-3 never went to trial and H-1 proved its accuracy but was kept from making the required long voyage.

Yet, in the Board's defense, they knew that to be successful the sea watch, or chronometer as it was now being called, had to be reproducible in numbers and prices that made it practical for the Royal Navy and Merchant Marine to use. The first reproduction of H-4 was by Larcum Kendall (1721–95), a watchmaker who was not only familiar with H-4 but helped build parts of it. However, it took Kendall 2½ years to make the reproduction.

K-1, as it was called, performed so well for Captain James Cook, the world's most able and renowned navigator, that he proclaimed it as "our trusty friend, the watch" on its sea trial in 1772. But the Board of Longitude failed to award the Longitude Prize, simply because the watch was not reproducible on a scale and price that the Royal Navy could afford for each of its vessels. Harrison made his second longitude watch, H-5, while Kendall was making K-1, to prove that H-4 was reproducible, but the Board still resisted.

At age 80, John Harrison, with his son William, appealed to King George III and, with the King's support, Parliament awarded John Harrison £8,750, making a total of £23,065 that Harrison had received in research grants over the course of his 40-year pursuit of the prize. In actuality, Harrison received more than the £20,000 award, but he never had the distinction of being awarded the prize and no one else did either. Harrison proved, with a lifetime commitment, that a portable timekeeper could and did solve the three-century-old quest for longitude. He introduced numerous technological advances into precision watchmaking; in fact we could say he pioneered the field of precision timekeeping. But, curiously enough, none of his chronometer designs were ever used again.

...AND HIS LEGACY

Out of efficiency, Harrison used the leading watchmakers of his time—all associated with George Graham's shop in London build portions of his later timepieces. And it was for John Jeffery, Larcum Kendall and Thomas Mudge, all either employees or apprentices of George Graham, to carry the torch of chronometer making.

The Board of Longitude has been depicted by historians as spineless, deceitful and arrogant, and John Harrison characterized as downtrodden, exploited and persecuted by the scientific community. But, to their credit the Board, although prejudiced toward an astronomical solution, did obviously feel that the intent of Parliament was to find a workable solution that the mariners of the world could afford and apply. And to their credit, the Board did grant substantial monies to support Harrison's research and development of the clock method of finding longitude to the tune of approximately £100 a year—a sum nearly equal to the annual salary received by the Astronomer Royal—for the first half of Harrison's career, and averaging £350 a year over his 40-year pursuit of the perfect marine timekeeper (not counting the £8,750 he received from Parliament).

Following directly upon the successes of John Harrison's timekeepers, and the fundamental truths that he applied, the newly established chronometer builders in England and their counterparts in France (Pierre LeRoy, Ferdinand Berthoud and Henry Sully) began producing accurate seagoing timepieces in numbers large enough and in times short enough for them to be usable by the Royal Navy and Merchant Marine.

BELOW
The Thomas Earnshaw marine chronometer, circa 1800. The movement is suspended in a set of gimbals—a system of rings that act like cradles or hammocks—to ensure that the instrument remains in a stable, horizontal position regardless of the pitching and rolling of the ship.

However, it was upon the invention of a form of new escapement by two young English chronometer makers that all subsequent chronometers would be made. John Arnold (1736–99) invented the pivoted detent escapement in 1782; Thomas Earnshaw (1749–1829) invented the spring detent escapement in 1783 and Arnold claimed that he did, also. The bitter debate as to who actually was the inventor continued in court and in public for years; and even today historians are not satisfied whether it was one or the other, or both independently. But the fact is that both of these men simplified the chronometer and perfected it to such a degree as to create the standard modern marine

ABOVE

A pocket chronometer by Thomas Earnshaw of London, 1800. Earnshaw was one of a half-dozen makers who brought the marine chronometer from its infancy to its perfection. One of his contributions was the spring detent escapement, which he also used in his highest quality pocket watches, such as the cased dial and movement pictured here.

chronometer that served the world virtually unchanged, right into the late twentieth century.

Finding longitude opened up a whole new era in empire building, colonization, commerce and economics. The improvements in navigation that the chronometer introduced allowed for safer and more direct passage across the oceans, creating greater intercontinental trade and opening up vast new markets. Knowledge of longitude created new vistas for discovery, not only of land and resources, but of new plant and animal life that greatly enhanced the knowledge of natural history. New cultures were found and populations emigrated and expanded. But the greatest contribution longitude made was to mapping. Land masses and coastal points could be mapped far more accurately than by the use of any astronomical method; and these correct maps led to all of the above gains in a far more rapid and safer mode than was ever dreamed of before the advent of the chronometer.

OUT OF CHAOS

Calls for a uniform national time standard surfaced periodically throughout the first half of the nineteenth century. Early reformers saw the standardization of time as both a practical way of organizing their work and an enlightened pathway toward progress. They approached the railroad managers with suggestions, eventually using Charles Dowd's idea of four time zones, each of 15 degrees of longitude equivalent to one hour as the sun moved across the sky. Dowd,

RAILROAD TIME

The Industrial Revolution brought about a general increase in wealth, and it brought watches within the reach of an ever-growing number of people. It also brought railroads to distribute the products of this manufacturing revolution.

Trains not only *left* a particular destination, but they *arrived* at one, too. And it was equally important for the trainmen, as well as the passengers and shippers, to know the time limits of both.

In the early days of train travel, the trains tried to operate within the numerous local time standards they encountered as they went from town to town. Each town had its own time, based on the local factory's noontime whistle, or the local town hall clock. This made scheduling from town to town nearly impossible. What evolved was a patchwork of schedules that was as hard to read as it was to understand and use.

In 1837 the newly invented telegraph was used on a London-based railway to send an instantaneous time signal to every point on the line. The effect was to establish a time standard for the railway's system. By the end of 1847 all of Britain was operating on a single railway time. Within a very short time all the rest of the European countries that had railroads adopted a single national hour of their own.

For the United States, Canada and Mexico, the countries were so vast that a single national hour would not do. Noontime in New York City would be 9 A.M. in California. Railroads and telegraph lines spanned the continent with speed. By 1830 only 23 miles of track had been laid; 35 years later there were over 93,000 miles. Telegraph lines spread even faster. In only 13 years from its 1860 introduction, 50,000 miles of telegraph wire was in operation.

Clocks were as vital to North America's railroads as were their locomotives. To avoid collisions on the same track, trainmen needed accurate watches and clocks synchronized to a standard time. But operating safety and punctuality were hampered severely by the multitude of regional and local times that had no logical relationship from one town to another, let alone from one state to another. Even though a railroad might operate on its own time standard to towns on its own route, the local people didn't necessarily do so; and, when the train used another system's tracks, safety was of crucial concern. By 1870 each railroad had its own time standard and there were more than 100 regional times in the United States.

The multiplicity of local times from area to area across the United States made the scheduling of accurate train services virtually impossible.

ABOVE
The high-quality and reliable Howard railroad chronometer, circa 1914. Edward Howard was the leading light and one of the founders of the Waltham family of watch companies. In 1902 the Howard name was purchased by the Keystone Watch Case Company. The name used on their watches was E. Howard Watch Co., Boston, USA, and this new company marketed only a complete watch, cased and timed in their own factory.

president of Temple Grove Seminary in Sarasota Springs, New York, campaigned for a standard railroad time system. Four North American science societies joined the campaign, led most notably by Cleveland Abbe, formerly of the Cincinnati Observatory and more lately the first official weather forecaster for the U.S. government.

At about the same time, railroad managers of competing lines began meeting to standardize their ever-expanding and interconnecting transportation network: standard track gauge, standard couplers, air brakes and signals and...standard time. Out of their twice-yearly meetings grew the General Time Convention and the Southern Railway Time Convention. W.F. Allen, secretary of the General Time Convention, is credited with putting into action the plan that the two conventions adopted in April 1883—a plan originally conceived by Charles Dowd. With vigorous promotion the idea was sold to businessmen, politicians, journalists and the North American public.

THE DAY OF TWO NOONS

On November 18, 1883, at noon on the 75th meridian of longitude, standard railroad time went into effect in the United States. Cities and towns across America had two noons that day, one on local time and one on standard time. Just as there was no public outcry for standardization, there was little objection, either. The American people quietly accepted and adapted their daily lives to it over a period of time. Towns and cities and states put their localities onto the railroad standard. There were some uncomfortable problems with some of the time-zone boundaries, but over the years the boundaries moved to accommodate residents living near larger commercial centers that happened to be just beyond the center's time-

RIGHT
An Illinois, size 18, approved railroad watch with 17 jewels. Dated to around 1899, the case is housed in a metal case with a locomotive engraved on the back. The Illinois Watch Company was set up in 1869 and sold to the Hamilton Watch Company in 1927.

zone boundaries. And there were some religious fundamentalists who saw the time change as interfering with "God's Time." But all in all, railroad time became national time and, curiously enough, 25 years later the U.S. Congress got around to adopting it officially as the national time standard.

A WORLD STANDARD

North America's standard time recognized Greenwich, England, as the prime meridian from which distance and time zones were measured. Sandford Fleming (1827–1915), chief engineer of the Inter-colonial Railway and later of the Canadian Pacific Railway, is credited with the idea of worldwide time zones. But it was the International Geographical Congress held in Antwerp, Belgium, in 1870 that organized international support for the idea. There, scientists met periodically to draw up a proposal for a world standard of time. In October of 1884, diplomats and specialists meeting at the International Meridian Conference in Washington, D.C. recommended that all nations establish their prime meridian at Greenwich Observatory outside London and count longitude east and west from this meridian, with each successive principal meridian of 15 degrees (or one hour's passage of the sun), making 24 one-hour time zones around the world. A universal day would begin at midnight in Greenwich. Although the conference had no authority to enforce its suggestions, the world gradually adopted the system almost universally. And, from that time onward, all systems, from scientific research to ocean shipping, airline travel, land transportation and communications, have used coordinated universal time based on midnight in Greenwich as the world's standard of time.

BELOW
The Old Greenwich Observatory. The prime meridian of the world extends in front of the building that houses the transit circle, the instrument upon which Greenwich Mean Time was based. Although the Royal Observatory left this location at Greenwich shortly before World War II, the Greenwich meridian still remains the point of zero longitude, the reference for determining Coordinated Universal Time.

TRAIN WRECK

The patchwork pattern of local times and the seemingly incomprehensible timetables that accompanied them were eliminated when the railroads in North America adopted uniform standard time zones for their scheduling. Accurate, dependable timepieces were required to keep the trains running safely and on schedule. Telegraphs transmitted correct time signals to various stops along the railroad, and train personnel checked and set their watches frequently with these signals. But there was a large variety of clocks and watches of varying qualities used to keep the railroads' time and to keep the trains on schedule.

And then, on April 19, 1891, there was a disastrous head-on collision between two trains on the Lake Shore and Michigan Southern Railway at Kipton, Ohio. One engineer's watch had stopped for four minutes and then started again on its own. The engineer didn't realize what had happened. He depended on his watch to tell him when he had to be off the main line and onto a siding so that an oncoming train could continue through. He thought he still had a window of four minutes. Both engineers and nine mail clerks were killed.

Government investigators and railroad officials asked Webb C. Ball (1874–*c.*1918) of Cleveland, a watchmaker and general time inspector for more than 125,000 miles of railroad in the U.S., Canada and Mexico, to organize a system for standardizing and inspecting the timepieces used on the railroads. Ball's recommendation, which was adopted by all the great railways, included a number of technical features vital to the

A nineteenth-century painting by A. Povart, depicting a railway accident on the Versailles-Bellevue line, France, May 8, 1842. Inaccurate timepieces led to errors in scheduling train traffic, which led, in many instances, to disastrous consequences.

The Hamilton model 992 railroad watch. Arguably the most poplar and most widely used railroad quality watch in America, it was made by the Hamilton Watch Company of Lancaster, Pennsylvania.

The 992 was made from 1902–39; its successor, the 992B, was made from 1940–69. Both models incorporated the finest materials and were made to the most exacting standards of any watch made from entirely machine-made interchangeable components.

precise functioning of a quality watch, as well as features necessary for the human handling of the watch. They had to be 18- or 16-size (based on the old Lancashire system of measuring watches—*see* Size, page 170) and have no fewer than 17 functioning wheel-bearing jewels. The balance had to be adjusted to run within close tolerances in five positions in which the watch might be carried (eventually a sixth position, with the winding stem lying in a downward position, was added). The hairspring had to be of the Breguet over-coil style to reduce variations in balance oscillation over the course of daily running; and it had to be controlled with a patented regulator. The balance wheel had to be adjusted to oscillate consistently in temperatures between 30 and 90 degrees Fahrenheit. The watch's hands had to be set with a pull-out lever mechanism rather than a stem, because a stem could accidentally pull into setting position in the pocket and the hand could accidentally move. The dial had to have a plain white background, with Arabic numerals and large black hands, and with the 12 o'clock positioned at the stem side of the watch. The watches were to be made by any American manufacturer who had a registered trademark.

Webb Ball also set up a time service inspection system. With offices in Cleveland, Chicago and San Francisco, he had inspectors for all the great railway systems, and he used local watchmakers as local time inspectors. Employees were required to have their watches inspected every two weeks, and a "loaner" was given for use while any necessary repairs and adjustments were made to their watch. The watches had to keep time within 30 seconds or less throughout the two-week period between inspections. Each employee carried a card showing the complete inspection record and performance of his watch, signed by the inspector. Watches were originally required to be serviced, cleaned and oiled once each year until newer oils and lubricants were developed that would allow a watch to perform reliably for a two-year service period.

CHAPTER 8

THE QUARTZ REVOLUTION

"One night I dreamed I was locked within my father's watch/With Ptolemy and twenty-one ruby stars/Mounted on spheres and the Primum Mobile/Coiled and gleaming to the end of space/And the notched spheres eating each other's rinds/To the last tooth of time, and the case closed."

JOHN CIARDI, *MY FATHER'S WATCH*

ABOVE
Flamsteed House and courtyard, site of the Old Royal Observatory at Greenwich. Half a century before the entire world began getting time signals from Greenwich, observatory officials dropped a time ball every day at 1p.m., so that ships on the River Thames could set their chronometers. In 1852 a master clock was installed to drop the time ball, send time signals via telegraph lines to the railways and display the time on a clock outside at the gate to Flamsteed House.

The era of precision timekeeping that started with John Harrison and grew under the skillful hands of the great chronometer makers of England and France experienced a major advance in 1899 when Charles Édouard Guillaume (1861–1938) introduced a new material known as *invar*. An alloy of iron, nickel and manganese, invar has a very low coefficient of expansion—meaning that under conditions of extreme heat or cold, it neither expanded nor contracted as other metals did. It very nearly overcame the problems of compensating for temperature variations and became an ideal material for balance wheels and pendulums. A similar material, called *elinvar*, was developed by Guillaume in 1913 for balance springs. These materials paved the way for even greater advances in precision accuracy. In 1920 Guillaume was awarded the Nobel Prize for physics for this discovery.

Between 1918 and 1939 two major advances were introduced that began the move away from a purely mechanical means of precision timekeeping. William Hamilton Shortt invented the free-pendulum clock in 1921. Shortt's timekeeping system used two clocks, a master and a slave, linked together electrically. The master clock's free pendulum received an impulse every 30 seconds from the slave clock and in turn transmitted a synchronous signal back to the slave. In 1925 it was introduced as the standard timekeeper for the Greenwich Observatory. This timekeeper was capable of providing an accuracy of one-tenth of a second per year.

In 1928 Warren A. Marrison (1896–1980) of the Bell Telephone Laboratories in New York developed the first quartz crystal clock. This invention replaced the escapement and pendulum of earlier clocks and created an entirely new concept in time measurement. By the end of World War II quartz clocks could sustain an accuracy of about 1 second's variance in 30 years.

ABOVE
The Junghans "Astro-Chron" quartz desk clock, 1967. This is one of the first domestic quartz clocks, first introduced in the late 1920s, but at that time its use was confined to observatories, laboratories and time standards. Advances in electronics and battery technology brought the quartz clock into the home and the quartz watch onto the wrist in the late 1960s.

In 1948 the atomic clock was developed for use as a frequency standard at the National Bureau of Standards in Washington, D.C. The atomic clock has not actually displaced the precision quartz clock, but it serves as the ultimate frequency standard to check and adjust the accuracy of the quartz clock.

The entire history of watches has evolved around a continual quest for improved accuracy and reduced size. Accuracy was a function of both scientific innovation and technical application, while smaller and smaller sizes were a measure of the craftsman's skill and inventiveness. By the 1960s, scientists and watchmakers had brought the mechanical watch nearly to the zenith of its potential in both precision and compact size. The only progress possible was to move to other forms of timekeeping for the masses.

A NEW REVOLUTION BREWS

As the Swiss watchmaking companies were cranking out close to 34 million mechanical watches a year—four times as many as Switzerland's nearest rival, the United States—a quiet revolution in watchmaking was beginning to brew. The year was 1952, and through the collaboration of the Fred Lip Company of Besançon, France, and the engineers at the Elgin Watch Company in the United States, the first battery-powered wristwatch was born. It was a "laboratory model" that was never marketed, but shortly afterward, in 1957, the Hamilton Watch Company put a battery-powered watch into production, and the Hamilton model 500 became the first commercially successful electric watch.

A NEW DEFINITION OF TIME

The advent of the atomic clock changed our perception of time. Subsequent experiments led to the introduction of the cesium 133 atom as a frequency standard. As cesium atomic clocks were improved, they became capable of maintaining an accuracy rate of about one second in 3,000 years. With this accuracy, astronomers realized that our earth clock, which we depended upon for so long, is slowing down by about one second a year. The stars, which science depended upon as the standard for timekeeping accuracy until the advent of the atomic clock, are actually drifting in direct relation to the earth.

So, in 1967, the atomic second was internationally adopted as the fundamental unit of time. The standard time used throughout the world is still based upon the Greenwich Meridian but it is now known as Coordinated Universal Time. The accuracy of Coordinated Universal Time is monitored and maintained by the Bureau International de l'Heure in Paris, by coordinating the rates of 80 atomic clocks in 24 participating countries.

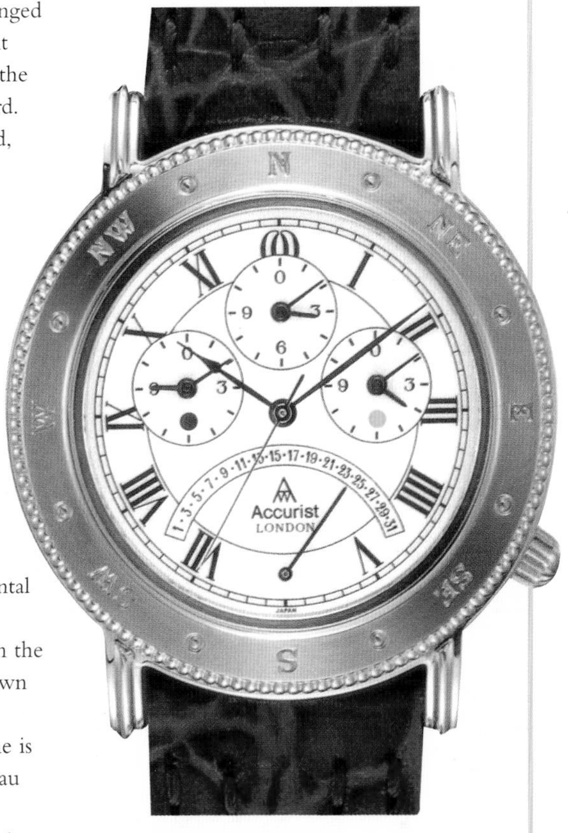

A multidial wristwatch that shows the date and the time on four dials in four different time zones.

The cesium atomic clock made in England by the National Physical Laboratory of Middlesex, 1955. The timekeeping depends on the vibrating cesium atom and is, in itself, a frequency standard and not really a clock that indicates time.

THE ELECTRICAL WATCH

Hamilton placed electrical coils on the balance wheel of the watch and, as the balance wheel oscillated, it passed between permanent magnets, making the balance into an oscillatory motor that drove the gear train of the watch. The secret of the Hamilton electric watch was in the mechanical switching system. A continuous on/off switching action allowed electrical current to impulse the balance, turning the balance into an oscillating motor, yet permitting the free oscillation of the balance wheel so that it could be used as a frequency standard. Advances in electronics introduced diodes and transistors into the electric watch's circuitry, permitting electronic switching of the oscillator/motor, and eliminating the mechanical switching system of the first electric watches.

ABOVE
Hamilton electric 1958 "Ventura" model—the first battery-powered wristwatch. Though superseded by the tuning fork and then the quartz wristwatch, the Hamilton electric watches were a significant advancement in the development of electronic timekeeping. Timekeeping was still controlled by a balance wheel, but the balance was kept in motion by electro-magnets instead of by a mainspring.

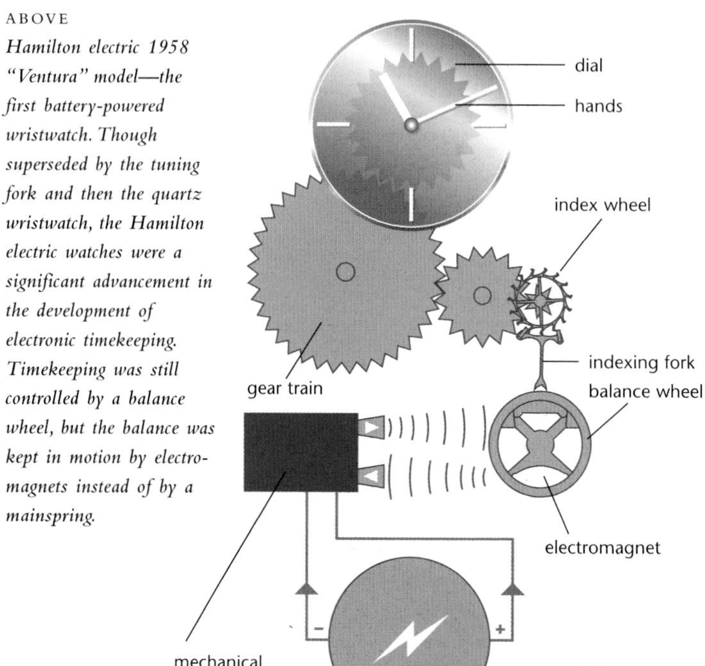

LEFT
An electrical watch uses a battery-powered electromagnet to energize the balance wheel. To maintain a stable oscillating frequency, the balance must be energized briefly and then the energy must be switched off to let the balance complete its swing before the energy is switched on again to energize its next swing. The balance swings four, five or even six times a second, driving the gear wheel train that turns the hands.

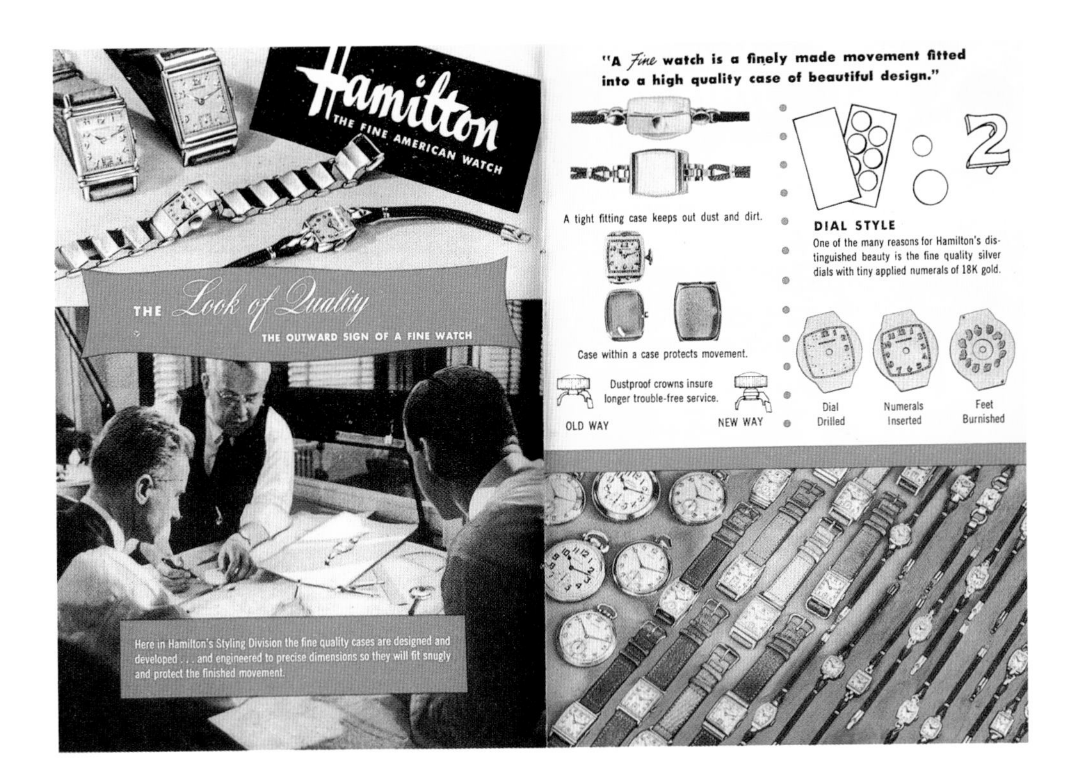

Every timekeeper, from the first mechanical clock in the fourteenth century to the atomic clock of today, requires:

- an oscillator or frequency standard that swings or oscillates back and forth in equal intervals
- a power source that energizes the swings
- an indexing system that counts the swings or oscillations.

The pendulum of the weight-driven clock performs the same operation as the balance wheel of the spring-powered watch, the vibrating quartz crystal of the quartz clock and the vibrating atom of the atomic clock. The faster the oscillations, the less inclined they are to be disturbed by outside forces, and therefore the more stable is their frequency.

THEY BECAME ACCURATE

For 400 years the standard of the portable timepiece was the balance wheel—even for the electronic watch. But in a complete departure from tradition, the Bulova Watch Company introduced a vibrating tuning fork as the frequency standard in 1960. The technique was the invention of Max Hetzel, a Swiss engineer, who offered it first to the major Swiss watchmakers. When the Swiss showed no interest, he took

ABOVE

A 1940s original brochure from the Hamilton Watch Company of Lancaster, Pennsylvania, explains that a fine watch is more than a well-made, highly accurate movement. To protect the movement from ambient moisture and contamination, a fine watch must have an exquisite case with a snug-fitting back cover and a crystal (or glass window.) The crown (or winding knob) must also fit tightly against the case. Above all, a fine watch must have the "Look of Quality"—a beautiful design suited to the style of the day.

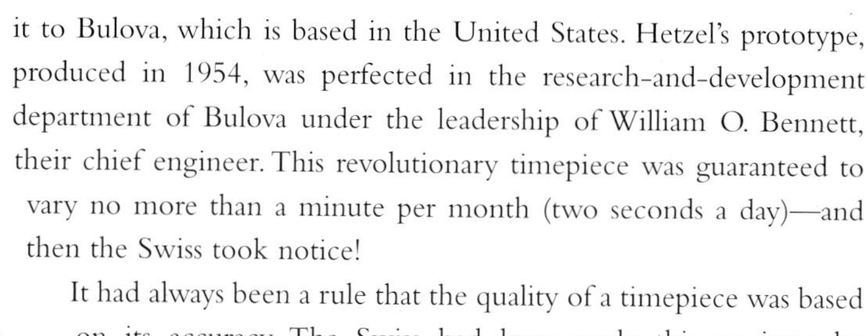

it to Bulova, which is based in the United States. Hetzel's prototype, produced in 1954, was perfected in the research-and-development department of Bulova under the leadership of William O. Bennett, their chief engineer. This revolutionary timepiece was guaranteed to vary no more than a minute per month (two seconds a day)—and then the Swiss took notice!

It had always been a rule that the quality of a timepiece was based on its accuracy. The Swiss had long made this an issue by certifying their precision watches through Swiss observatory testing. When Bulova offered a better product—i.e., one that was more accurate than most of the Swiss companies could produce—the Swiss and the Japanese reacted by developing an even better oscillating standard than the tuning fork for the wristwatch: the vibrating quartz crystal.

The history of the quartz crystal goes back to 1880, when Pierre Curie (1859–1906) and his brother Jacques found that certain crystals emit electrical charges when they are stressed (this is the same Pierre Curie who later married Marie and together with her discovered radium and radioactivity). The piezoelectric effect, as it is called, was first applied to submarine detectors in 1916. Walter Cody of Wesleyan University, Connecticut, applied the quartz crystal to frequency-control units in radio broadcasting; and in 1928 Warren Marrison applied it to the clock (see page 145).

ABOVE

The Bulova Accutron wristwatch uses an electronically driven tuning fork as an oscillator or frequency standard. The Accutron ushered in a new era in watchmaking, changing the way we viewed accuracy and, although short-lived in itself, it set the stage for a more accurate frequency standard to come.

It was nearly 30 years later that the quartz crystal was applied to wristwatches by both Hattori Seiko of Japan and the Swiss Center for Electronic Horology, both working and developing the quartz wristwatch independently of each other in 1967. Seiko was the first to market its version in 1969, it was followed quickly by the Swiss in 1970. The watch used an electronic circuit to process the 32,768 vibrations per second of the energized quartz crystal into one-second impulses of a motor. It was called an *analogue* quartz watch because the motor drove a set of gear wheels that turned familiar hands around a conventional watch dial to display the time.

At the time that the analogue quartz watch was emerging from Japan and Switzerland, the Hamilton Watch Company of the United States debuted the solid-state quartz watch. In this watch there were no moving parts at all. It used the same quartz technology and integrated circuit as the analogue, but instead of gears and hands, the module (you couldn't call it a "movement" any more) energized a series of light-emitting diodes that showed the time in a digital format. The LED display, as it was called, quickly went out of favor because it was difficult to see and used far more energy than the watch batteries could supply over a reasonable period of time. However, another form of digital display soon followed. Called a liquid-crystal display (LCD), it used a phenomenon that was first observed in 1888 in which particles in substances called liquid crystals could be rearranged into various configurations by applying an electrical current.

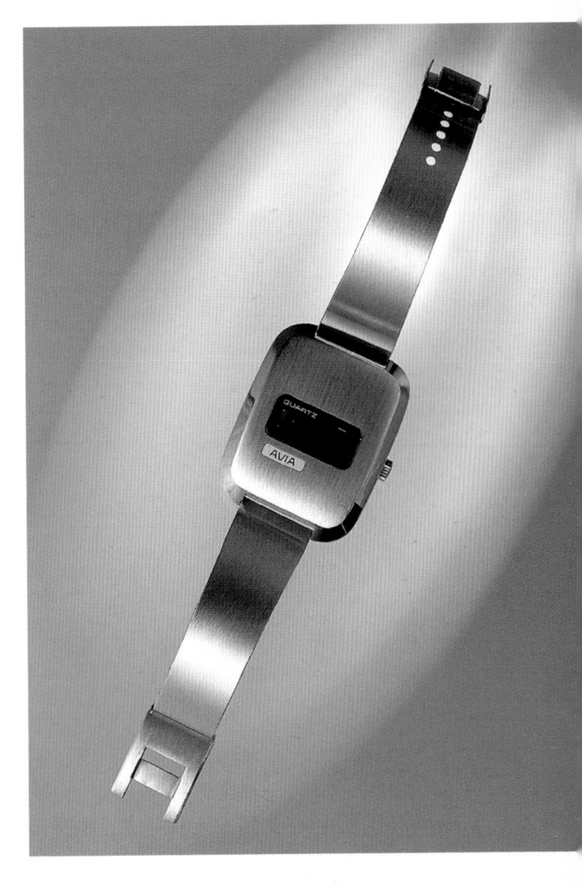

AND THEN THEY BECAME CHEAP

New materials, advances in miniaturization and microchip technology, and the fact that quartz watches use so few component parts combined with economies of scale in manufacturing to bring the price of these watches from a $1,000 retail in the early 1970s to less than $10 apiece today—if they are housed in a plastic case. The LCD, LED and analogue watches remain popular with the younger market because a digital watch can be produced even more cheaply than an analogue—and it can be put into a quality metal case and sold for under $50.

The analogue display remains popular with the twentysomething and upward age bracket, not so much because of the traditional dial and hands, but more because the traditional dial is still more quickly and easily used. The hands give you a spatial relationship to both the current time and to some past or future point of time on the dial. With the digital display, the mind has to calculate the relationship that the analogue gives at a glance.

Today's lower-priced watches are literally throwaway, disposable items. The mid-range quartz watches (most analogue types) are housed in stainless steel and gold-plated cases. The higher-end quartz analogue watches are housed in karat-gold and stainless steel cases, many with gemstones. And almost all of the mid- to high-end watches are also protected against moisture and dust, giving a much longer life and a much longer-lasting look than the low-end versions.

The microchip has made a myriad features possible on a watch. Not only are calendar features, along with stopwatch, timer and chronograph functions commonplace, but such additional features as calculators, radio-controlled time-setting, voice features, remote messaging, reminder memos, address and telephone databases, blood-

ABOVE
The advent of the solid-state watch in the early 1970s sent shock waves through the watch industry, tolling a death knell to traditional watches, portending the demise of the watch repair industry and threatening to move the manufacture of watches to the electronics industry. Ironically, none of this happened, but at the time it all seemed inevitable.

HOW THE MECHANICAL WATCH WORKS

The mechanical watch uses a mainspring to drive a set of gears. The last wheel in this train of gears is called the escape wheel. The escape wheel is a specialized wheel with angled teeth that performs two functions: (a) it drives a fork-shaped lever that changes the circular motion of the gears into an angular, back-and-forth motion; and (b) it counts this back-and-forth motion.

The pallet fork, as this lever is called, has two angled jewels, one on each of the two tines of the fork. As an escape-wheel tooth lifts one tine, the tail of the fork impulses the balance wheel, causing it to turn, or oscillate, in one direction. The tail of the fork immediately disengages from the balance, allowing the balance wheel to continue its revolution freely, i.e., not connected to the gear train or fork, providing a still uninhibited frequency.

A nineteeth-century tourbillon watch by Hunt & Roskell, London. The balance wheel and escapement are carried in a tourbillon cage thus obviating timekeeping problems caused by gravity.

At the end of its revolution, the hairspring reverses the balance wheel's direction and returns it to its starting point, where it connects again with the pallet fork tail. The escape wheel lifts the other tine of the pallet fork, impulsing the balance wheel so that the balance can continue its trip in this opposite direction. Upon impulsing the balance, the fork immediately disconnects again, until it reaches the end of its revolution and the hairspring reverses the balance's direction once again to return it to its starting point.

In the standard balance wheel and hairspring, this back-and-forth rotation, or oscillation, of the balance wheel happens five times each second. The escape wheel counts these five swings each second and blocks and releases the gear train five times each second in a rhythm that is even, smooth and constant. The gear wheels are calculated so that, when the escape wheel releases the gear wheels, the fourth wheel from the main wheel revolves once each minute; the center wheel (second wheel) revolves once each hour and the main wheel (or first wheel) revolves once every three hours. The second hand is linked to the fourth wheel; the minute hand is on the center wheel; and the hour hand is linked to a small series of gears that multiply the one-hour turn of the center wheel into a 12-hour turn of the hour wheel.

The escape wheel and pallet fork of this watch are together called the *detached lever escapement*. This is and has been the most commonly used escapement for quality watches for the past 150 years, at least. It was invented by Thomas Mudge (1715–94) of London in 1769. It was improved by Peter Litherland (*d.*1805) of Liverpool in 1791 and perfected by Edward Massey (1772–1852) of

London in the 1820s. The Swiss took up Massey's detached lever escapement and put it into commercial production in the 1840s. Today this same escapement is in most of the mechanical watches made, in multiples of millions each year.

The balance wheel and its hairspring govern the speed at which the gear wheels, with their indicator hands, are released. The back-and-forth motion of the balance wheel is called its *oscillation* and the speed at which it oscillates is called its *frequency of oscillation* or *frequency*. The standard frequency of today is 18,000 oscillations per hour (five beats per second); the standard of one hundred and some years ago was 14,400 oscillations per hour (four beats per second). Prior to that it varied by manufacturer. Some makers even made faster frequencies of oscillation, giving perhaps an even greater degree of accuracy to their watches. But bear in mind that accuracy is not only a function of oscillation frequency; it also depends on the lack of friction, the smoothness of the gear train's power transmission, the poise or balance of the balance wheel and the compensations for changes in temperature, barometric pressure and, as the mainspring runs down from fully wound to unwound, varying degrees of a balance's arc of oscillation.

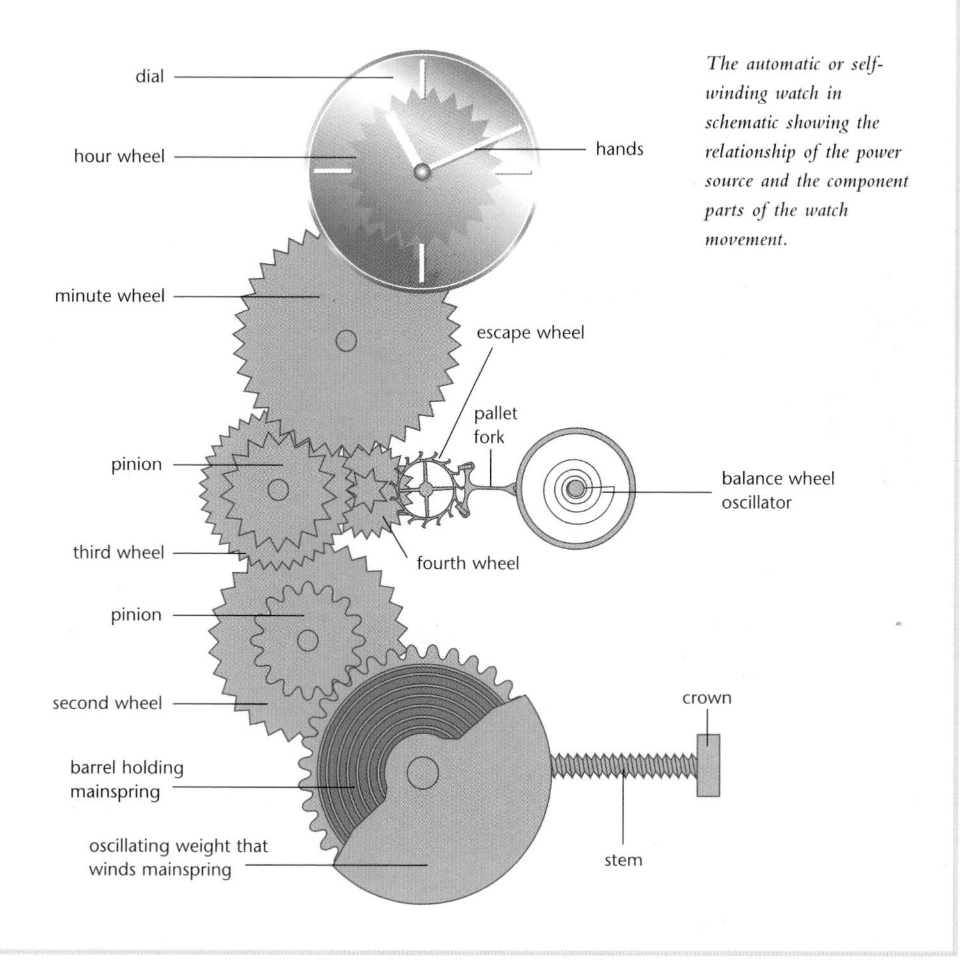

The automatic or self-winding watch in schematic showing the relationship of the power source and the component parts of the watch movement.

pressure monitoring, ultraviolet sensors and even radiation sensors are now available with the wristwatch. Even using a moving weight put into motion by the movement of the wearer's wrist winds a micro-generator to generate electricity, replacing the battery as the power source for some quartz watches. Called an "autogenerating" or "self-charging" quartz watch, the old concept of using a swinging weight, put into motion by the movement of the wearer's body, was invented by Abraham-Louis Perrelet, of Neuchâtel (1729–1826). As we peer into the future, watches appear to blur with computers, radios and televisions as new discoveries find additional applications to the organizing of our lives.

ACCURACY IS NO LONGER AN ISSUE

After 350 years, the wearable, portable timepiece is available and affordable to anyone who wants to have accurate time at a glance—even price is not an issue. The quartz revolution not only changed the watch; it had a revolutionary impact on those who make watches. As the electronic watch developed in the 1960s and 1970s, Swiss manufacturers, having helped pioneer the quartz watch, felt it would be a fad. They had the technology and they had the ability to produce these new timepieces, but they just didn't have the desire. They held on to the era that they were comfortable in: that of the mechanical watch. And they lost market ground to the U.S. manufacturers and then to the Japanese and Hong Kong manufacturers, who now produce most of the world's watches.

The Swiss still produce huge numbers of quartz watch movements on a par with all others in the world. Their unique system of *ébauche* production still gives them a large share of the unfinished movements that are sold and cased around the world. But they have lost a large share of the market for finished, cased watches. Just like the British manufacturers (when their production was surpassed by the Swiss in the early 1800s), the Swiss manufacturers have taken refuge

ABOVE AND RIGHT
Modern wristwatches by Longines Watch Company of Switzerland, with bracelet-style cases and bands, making them a fashion accessory as well as a timepiece for men and women.

HOW THE ELECTRONIC WATCH WORKS

The electrical watch is a broad category of watch that includes the *balance-wheel electric* and the *balance-wheel electronic* watch, as well as tuning-fork and quartz watches. They all function on similar principles, the only differences being the oscillator and the switching/counting system used.

Electrical watches use the electrically powered oscillator to drive their gears. The *electric* balance-wheel watch uses an oscillating balance wheel that is energized by a battery power source to drive the gear train five times each second. Just like the mechanical watch, the electric watch depends on the frequency of its balance to determine the speed of the gears; and, like the mechanical balance wheel, the electric version is impulsed and then released five times each second. But the electric watch uses electromechanical switches to do the on-and-off switching of the balance wheel. An index wheel, instead of an escape wheel, does the counting of the oscillations and turns the gear train at the appropriate frequency so that the center wheel turns once each hour.

In the *electronic* balance-wheel watch, the switching and counting of the oscillating balance wheel is done with electronic diodes and transistors, and so we get the adjective "electronic" rather than "electric."

The *tuning-fork* electronic watch and the electronic *quartz* watch use electronic switching and counting of the oscillator, just as the other electronic watches do, the exception being the type of oscillator used: the vibrating tuning fork versus the vibrating quartz crystal.

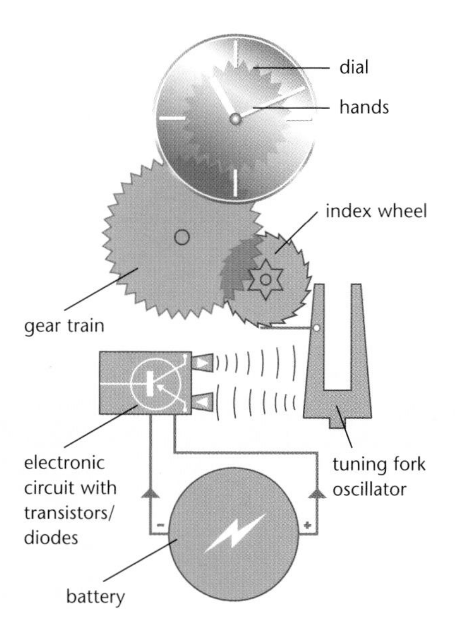

How the electronic watch with tuning fork works (See Accutron, page 150.)

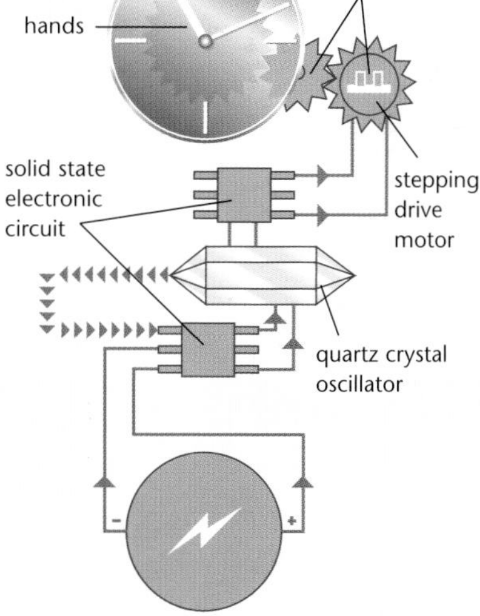

How the quartz watch works.

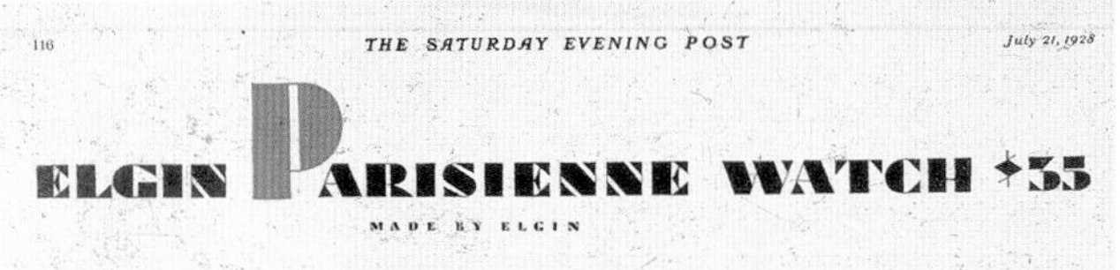

ABOVE

The Parisienne series of ladies' wristwatches by Elgin Watch Company of Elgin, Illinois. The cases of these watches were designed by leading Parisian designers using enamel inlaid on the top of (often) gold-filled cases. Made between 1927 and 1931, these watches were tied to the concepts of prestige, artistry and luxury.

in the higher-priced, luxury-watch market. Their production numbers are much lower than those of the Far East, but the value per watch is much higher. Switzerland still produces more mechanical watches than any other producing country, but that production is considerably down from what it was at it's peak. It is buoyed, however, by a recently renewed interest in the mechanical timepiece, tied most likely to the concepts of prestige, artistry and luxury that the Swiss marketers are very successfully selling to the world (despite this being an era of conspicuous consumption and throwaway commodities).

The mechanical clock and watch will probably one day go the way of the sundial and clepsydra, as will probably the quartz watches, as new technologies develop that we cannot even foresee at this time. But the legacy, the beauty and the artistry will remain. There will always be a niche for those who appreciate these wonders of the mechanical arts. There will always be certain craftsmen who seek the ultimate expression of their skills by creating samples of objects where science, art and a creative mind and skillful hand all come together in wonderful miniature machines that measure the time.

THE MECHANICAL WATCH IN THE QUARTZ AGE

A mechanical timepiece movement is really something to see. There is symmetry in the layout of the movement; there is balance and grace to the flow of its design. The plates and bridges are gilded and highly finished—swirled, decorated, chased and damascened (see glossary, page 164.) Every edge is beveled, sharp and crisp, and highly polished. The red jewels glisten in their gold settings. The screws are heated to

THE BATTERY—THE POWER BEHIND THE ELECTRIC WATCH

The electronic wristwatch depended on a small self-contained power source to make it work. In the early 1940s, Samuel Rubin created a self-contained, mercury energy cell that was manufactured by P.R. Mallory & Company (now Duracell, Inc.). Rubin's cell was small, but was later reduced to fit into heart pacemakers and hearing aids. This type of mercury power cell was used by Elgin Labs to energize their newly created laboratory model of an electric watch in 1952. When Hamilton needed a power cell for their commercially produced Hamilton Electric Watch, Union Carbide designed a self-contained, miniature carbon-zinc cell. Shortly afterward, a silver-oxide cell was produced which is still used in quartz watches today.

The term *energy cell* is the proper term for a unit of stored energy. A battery is actually a series of individual cells. Common parlance, however, has blurred the distinction in the public's mind and so *watch battery* is the term commonly used today.

The watch cell or battery is available in a broad range of sizes to accommodate the tastes of watch-case designers. Larger cells of course store more energy than smaller ones, but fashion rules, and people who want the smaller, thinner watches simply have their batteries changed more often.

Watch energy cell technology has evolved right along with the electronic watch. A small cell can produce more energy today than it could in years past, and for a much longer period of time. Different chemical systems for producing electrical energy have come to light. Mercury, producing 1.35 volts of electricity, is not used in the United States because it is a potential environmental hazard, the silver-oxide system produces 1.55 volts and is the most common cell used. Lithium cells, providing 3 volts in a wafer-thin package, lend themselves to watches with high-energy demands and with thinner cases. Watches that evolved to run on extremely low levels of energy can now be matched with cells that can store their energy longer without breaking down..

One of two pioneering quartz watch movements, this one is the Beta 21, designed by the Swiss Center of Electronic Horology (CEH) and manufactured by E.T.A S.A. of Grenchen, Switzerland. It was introduced as a prototype in 1967 and marketed in 1970.

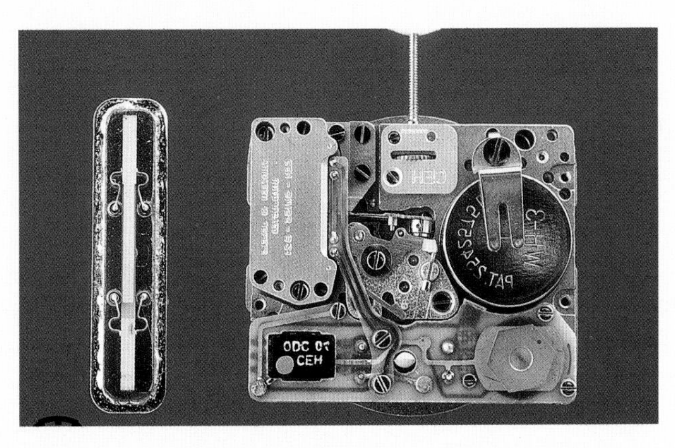

THE QUARTZ REVOLUTION AND THE REPAIR TECHNICIAN

The quartz revolution not only affected the manufacturing and marketing of watches on a world scale: it impacted the repair industry, as well. All of a sudden the repairman was confronted with an instrument that he didn't understand and a technology that was totally foreign to anything he had encountered before. The media proclaimed the demise of the watch repairman because these new watches wouldn't need servicing except for an occasional battery; and the old mechanical watch would be abandoned in favor of the new.

This was in the mid-1970s and the American Watchmakers-Clockmakers Institute (AWI) responded with a massive re-education effort. Sending their top instructors to U.S.-based quartz-watch manufacturers and to Swiss-based electronic-watch manufacturers, AWI learned the ins and outs of electronics sufficient for the watch repairman, devised repair techniques, and began retraining its 10,000-plus watch-repairman members for the new technology. AWI sent out teams of instructors throughout the U.S. and Canada, and over the next two decades presented

Watchmaking students in the 1950s studying the principles and functions of the detached lever escapement. In watchmaking schools, instructors use a lecture/discussion format for theory and practice; and then applied practical application in the watch laboratory.

a deep, rich electric blue. And there is more—much, much more even, than the artistry of its physical body: there is a life and a grace to the mechanical movement that transcends the fact that it is but a mere machine.

The balance wheel swings in an uninterrupted, untiring motion, to and fro; the hairspring moves in and out, breathing with the beat of the balance; the pallet fork ticks off seconds as it counts the rhythm of the balance wheel, releasing the power of the slowly moving train of wheels, so as to mark a day's passing. Even the audible rhythmic tick itself tells you the watch is alive, delighting the listening ear with a quiet music all its own.

Still, someone who knows nothing about the mechanics of a timepiece can appreciate its artistry and beauty. To that end, many of the finest watch movements are

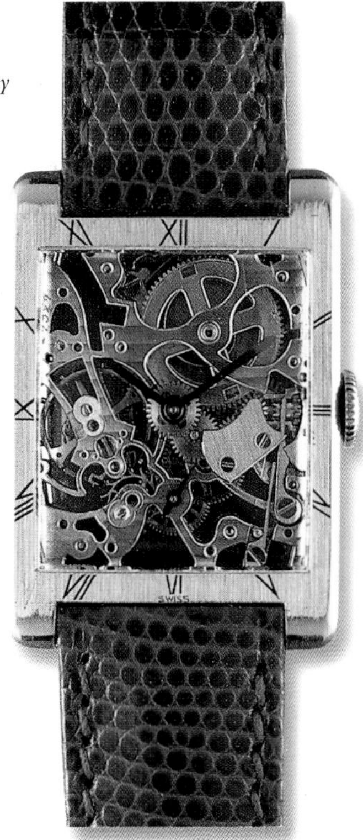

RIGHT
Reportedly the first skeletonized wristwatch by Audemars Piquet in 1937. Skeletonizing refers to the removal of much of the plates of the watch movement to reveal the artistry, beauty and rhythmic motion of the inner working of the mechanical watch.

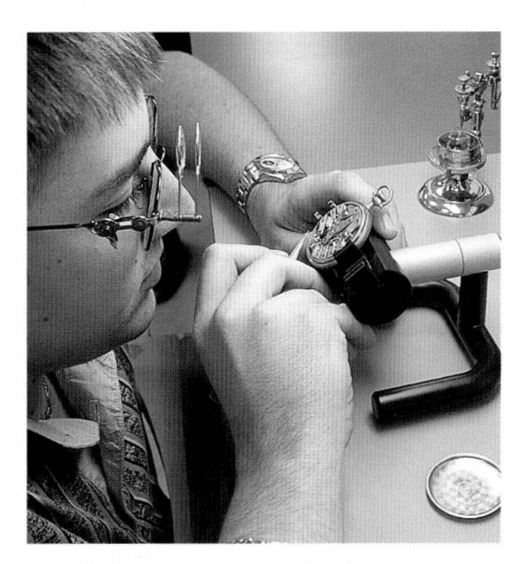

Watchmaking students in the state of the art laboratory at AWI's Academy of Watchmaking, Harrison, Ohio, preparing for a watch repair and watchmaking career.

hands-on bench courses to all who wanted to stay abreast of the rapidly and continually changing electronic-watch technology.

The effort worked: members and nonmembers alike learned that these electronic timepieces would need periodic servicing and that they could be serviced and repaired on an economical scale with a minimum of specialized equipment.

AWI also learned that mechanical-watch owners would not abandon their mechanical watches, either. They were too personal an item: they were gifts or possessions of loved ones and ancestors. Even today, the repair of old and antique timepieces is a growing field, and with the millions upon millions of new high-end and complicated mechanical watches coming on to the market each year, the prospects of continuing service and repair will run far into the future—for those who take the time to gain the knowledge and master the skills of the watchmaker.

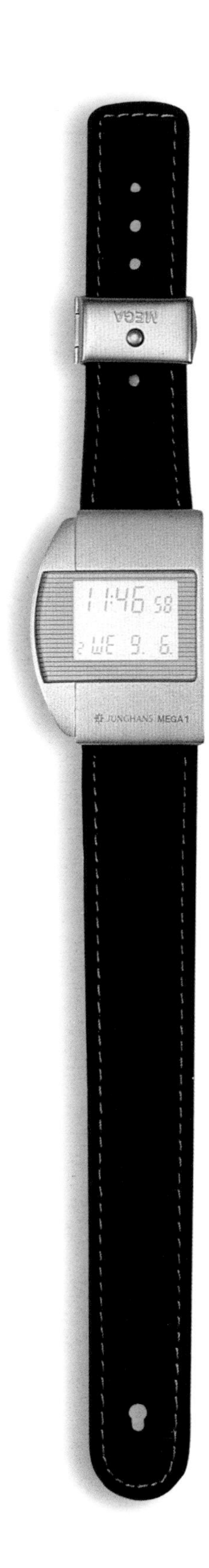

"skeletonized" and encased in see-through cases to reveal the inner workings and their symphony of motion for the pleasure of the watch's owner.

IT'S HARD TO LOVE A QUARTZ TIMEPIECE

The early quartz timepiece movement—if you can call it a movement, since there is little actual movement to its component parts—was a circuit board with transistors, diodes and other electronic components soldered onto circuit paths. This is especially so with the solid-state timepieces: they were a product of the electronics industry. Even the analogue timepieces were an expanded jumble of electronic components surrounding a train of five wheels under a small plate or bridge in the center of the movement. There was no beauty and grace, no artistry of design, no rhythm of motion and ... no living tick.

Today the Swiss analogue quartz movements are, in many cases, still made of a metal base, plated gold or silver, with pretty blue circuit boards and contrasting silver circuit paths—still an electronic commodity, not far different from a miniaturized version of your personal computer's motherboard. But the Swiss, too, are following the lead of the Japanese quartz-movement makers. Efficiency, technology and lower and lower unit prices have driven the Japanese to find newer and cheaper methods to make the quartz watch movement. Specialized plastics that can be molded to exact dimensions have replaced the more expensive machining processes on metal. Machined brass wheels yielded their place to stamped metal wheels and then to molded plastic wheels, as plastics technology advanced to its highest degree in precision, miniaturized molding and manufacture.

The most common Japanese and Swiss quartz movements are now black engineered plastic. The highly refined circuitry is printed on a fine thin film of plastic, sandwiched between plastic plates that hold five plastic wheels and the movement's rotor, also plastic. When seen by the casual observer the world's most advanced timepieces are reduced in the mind of the uninitiated to, "Is that all there is to a watch?"

The blend of scientific achievement, manufacturing precision and electronic sophistication is represented in the quartz analogue watch—totally produced by machines that the mind of man has devised and the hand of man has created. But the *soul* of the watch is not there. The soul of the watch is in the living, breathing, ticking mechanical movement that has had the individual touch of a human hand, even when mostly made by machine. The quartz movement is precise, efficient and utilitarian. The fine mechanical movements of today still exhibit qualities that the handcrafted ones showed in the past: flowing design, beveled edges, polished plates, and surface etchings, engravings, and patterns still applied by human hand. Even the entirely machine-made mechanical movement exhibits a life of its own with its breathing hairspring and rhythmically beating balance wheel.

OPPOSITE
The Junghans Mega radio-controlled wristwatch, introduced in 1992. The radio-controlled watch, like the radio-controlled clock, uses a miniature receiving set connected to an antenna that picks up a time signal from the National Bureau of Standards in the United States and participating observatories around the world via satellite relay. The observatory signal automatically corrects the oscillations of the watch's quartz crystal, giving the watch or clock nearly the same accuracy as that of the observatory's time standard.

LEFT
One of two pioneering quartz watches—the Seiko Astron Model 35, designed and manufactured by Hattori Seiko of Japan, introduced as a prototype in 1967, and marketed in 1969.

GLOSSARY

Analogue display: A timepiece that uses hands and a dial to display the time.

Anchor escapement: An anchor-shaped piece of steel that rocks back and forth by the swinging motion of the *pendulum*. This back-and-forth rocking motion releases a tooth of the *escape wheel* with each swing of the pendulum. At the same time, the escape wheel *impulses* the pendulum through the anchor escapement to keep it swinging. Also called a *recoil escapement* because the motion of the anchor imparts a slight backward motion to the escape wheel. The anchor escapement is found in almost all floor clocks; it supplanted the *verge escapement* and made accurate clocks possible. Probably invented by Dr. Robert Hooke of London (1635–1703) and was first used by William Clement (1638–1704), a clockmaker of London, in 1671.

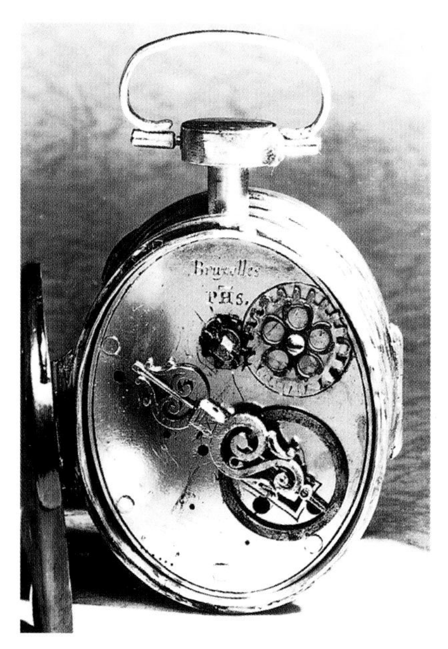

A Flemish Renaissance watch, circa 1590. This magnificent movement is in pristine condition and features an S-shaped pinned balance cock.

Ancient timepiece: A timepiece made when wheel teeth were laid out and shaped by hand, prior to the general use of the wheel cutting machine.

Antique timepiece: A timepiece made (roughly) between 1700 and 1900.

Arbor: A horological name for a wheel shaft or axle in a clock or watch. (See Stackfreed and *Fusée* box, page 36.)

Astronomical timepiece: A timepiece with a mechanism that shows various solar, lunar and stellar displays; e.g., sunrise and sunset times, phases of the moon, and *sidereal time*.

Automatic watch: A mechanical watch that uses the motion of the wearer's wrist to wind the *mainspring*. The mechanism that converts wrist movement into a winding motion uses a free-swinging oscillating weight and geared ratcheting system invented by Abraham-Louis Breguet (1747–1823). Also known as a self-winding watch.

Balance: See *Balance wheel*.

Balance bridge: A bracket supported on both ends and spanning the *arbor* of a *balance wheel*, providing a bearing point for the *wheel arbor*.

Balance cock: A bracket with one end supported and spanning the *arbor* of a *balance wheel*, providing a bearing point for the *wheel arbor*; with the other end being free.

Balance spring: A weak spiral spring whose inner end is attached to the axle or shaft of the *balance wheel*, and whose outer end is attached to the *plate* or *bridge* of the timekeeper. The spring winds and unwinds as the *balance* swings back and forth, giving a regular and consistent oscillation to the balance wheel. The *balance spring* is also known as a *hairspring*. (See Swiss Detached Lever Escapement, page 81.)

Balance staff: The *arbor* or axle upon which the *balance wheel* is mounted. (See Swiss Detached Lever Escapement, page 81.)

Balance wheel: A circular-shaped *controller* that

oscillates back and forth providing the timekeeping standard for a mechanical timepiece. Although properly known as a *balance*, the term *balance wheel* more aptly describes this wheel-shaped controller. (See The Principles of the Watch, page 93.)

Balance wheel controller: See *Balance wheel* and *Controller.*

Bearing hole: The hole in which a *wheel arbor* rotates. The bearing hole supports the *pivot* and provides a smooth surface in which the pivot can turn. (See The Principles of the Watch, page 93.)

Beveled edge: An angled corner or edge of a flat surface, inclined at any angle except 90 degrees. Beveling adds a pleasing aesthetic quality to the *bridges*, *plates* and flat parts of a timepiece *movement.*

Blanc: The French term for the basic frame of a watch *movement* without plating or engraving and often with a *mainspring barrel*, *fusée* and sometimes with the wheels and *pinions* roughly installed. A *blanc* with all its gear wheels installed and turning smoothly but without an *escapement* is called a *blanc roulant.*

Breguet overcoil: A *hairspring* whose outer quarter-turn is raised and curved over the body of the spring toward its center. An ordinary spiral spring does not keep good time because the spring becomes lopsided as it opens and closes (vibrates) with the swing of the *balance*. This overcoil, discovered by A-L Breguet (1747–1823) in about 1800, helps keep the *hairspring* more closely centered around the *balance staff* as it winds and unwinds (opens and closes).

Bridge: A metal bar that carries one or more *wheel pivot bearing holes* and is fixed or supported at both ends.

Cabinotier: Geneva watchmakers working in the small *cabinets* or shops lined up under the eaves of buildings in the Faubourg area of Geneva.

Calibre: The French term literally meaning *mold* or *model*. In *horology* a *calibre* means both the size and dimension of a watch *movement*; and the type of design or layout of a movement.

Carriage clock: A relatively small, portable clock, spring-driven with a *balance wheel controller*. The earliest ones were like very large pocket watches. Typical carriage clocks have a rectangular case with a handle on top. French styles typically have glass sides. Carriage clocks were so called because they could be carried from place to place, often in a carriage, before the advent of the pocket watch.

Case: The outer housing for a timepiece, used to protect the delicate mechanism within.

Chased: A decorated metal surface enhanced or augmented by hand engraving and carving.

A gold-cased pocket chronometer by John Arnold of London, 1780; with a spring detent escapement.

Chronograph: A timepiece and a *stopwatch* that both shows the time of day and measures short intervals of time.

Chronometer: A non-pendulum precision clock or watch. In Switzerland, no manufacturer can call a watch a chronometer unless it has obtained an official *rating certificate* from one of the recognized testing bureaus.

Circuit board: The platform or substrate that supports the electrical or electronic components of an electrical or electronic timepiece, such as transistors, diodes, capacitors and the electric circuits or pathways.

Clock: An instrument for measuring or indicating the passage of time. Generally referring to a stationary timepiece designed to be used in one place, either on the wall, floor, mantle, shelf or table.

Clockmaker: One whose occupation is making or repairing clocks. A clock repair technician is one who can replace defective component parts and adjust the clock's performance.

Clockwork: The *movement* or mechanism of a clock, consisting of a power source, a *gear train, escapement* and *controller*. Can also refer to an assemblage of clock-like gear wheels and *pinions.*

Co-axial escapement: A combination of a *lever* and a *detent escapement* with their two *escape wheels* revolving on a common axis to count the seconds. The co-axial escapement first appeared in the early twentieth century, but was refined by George Daniels to eliminate any sliding friction that earlier attempts had to contend with.

Complicated watch or clock: A timepiece with any additional mechanism not involved with telling the mean solar time or striking the hour; for example: a *repeating mechanism*, *chronograph*, calendar or astronomical works (each of which is called a separate *complication)*.

Complication: Any mechanism found on a watch or clock, other than the basic mechanism that shows the time of day. See *Complicated watch or clock*.

Controller: The part of a timekeeper that controls the timekeeping rate: either the *balance* or *pendulum*, or other *frequency oscillator*.

Corporation of Craftsmen: A body of craftsmen granted a charter legally recognizing them as a separate entity to engage in the pursuit and training of their craft. Such a body had its own rights, privileges and liabilities. Such corporations were called guilds whose jurisdictions extended to the legal boundaries of their guild areas, usually the city limits. The hierarchy consisted of the **Master**, who owned the shop and who did the training, the **Apprentice** who was undergoing the training, and the **Journeyman** who had completed the training but did not yet own his own shop.

Cottage system: A system of manufacture in which individuals specialized in the making of component parts for watches and/or clocks. The components then went to others specializing in the assembly, finishing and marketing of the completed timepiece.

Count wheel: A wheel that is the memory in the *striking mechanism* of a clock. It counts the number of hours to be struck on a bell or gong. Also known as a *locking plate* because it locks the *striking train* between hours, releasing it only at each hour for the duration of the striking count.

Crown: The knob or button on a watch used to wind the *mainspring* and/or set the hands.

Crown-verge wheel: See page 22.

Crown wheel: An *escape wheel* in the shape of a crown. Used on early *verge escapements*.

Cruciform watch: An early *form watch* whose case was in the shape of a cross or crucifix; many times the *movement* itself was also in the shape of a cross.

Cylinder escapement: A form of escapement that uses a half-cylinder recess on the *balance staff* to release the *escape wheel* teeth and to receive an *impulse* from the escape wheel. This escapement was perfected by George Graham in 1725.

Damascened: A decorated metal surface having a wavy pattern of inlay or etching.

Detent escapement: A *detached escapement* in which *the balance wheel* is impulsed in one direction only. During its swing the balance wheel moves aside a finger that releases one tooth of the *escape wheel*. This tooth gives the balance wheel a push in the direction it is going and is then locked. On the return swing of the balance a so-called passing spring prevents the escape wheel from being unlocked. Used by Thomas Mudge, Thomas Earnshaw and Pierre LeRoy in the early eighteenth century, it is also called a "chronometer escapement."

Dial: The gradated circular face of a timepiece that is used to display the time. It can be divided into twelve hourly segments and sixty one-minute segments to indicate the time through the use of moving indicator hands. A dial can have a number of **sub dials,** each with its own indicator hand to indicate other functions, such as seconds, lapsed time, day of the week and month of the year. An **aperture** or window in the dial can also show additional information in the form of figures, numbers or words.

Dial train: The *gear train* of a timepiece that is located under the *dial*. This train of gears reduces the one revolution per hour of the center wheel (with its minute hand) to one revolution per day of the hour wheel with its hour hand attached. Also known as the "motion works" because it puts the time display into motion.

Differential gearing: An arrangement of *gears* that permits the rotation of two shafts at different speeds.

A silver-cased **pendule de voyage** *or travel clock made by Breguet and Sons of Paris, 1822, with a calendar and half-quarter repeating mechanism.*

May be used in *equation timepieces*, astronomical works and up-down indicators that show how much wind remains on a *mainspring*.

Digital display: Displaying the time using numerals rather than indicator hands that point to numbers. In the past, some mechanical watches used numbers to display the time but the most common digital displays are the **LED**, or **light emitting diode**, which uses a system of very tiny pinpoints of light to form the numbers; or the **LCD**, or **liquid crystal display**, which uses an energized liquid that forms the numbers. The LED display must be turned on each time the time is to be seen; the LCD is on constantly for quick and easy viewing.

Ébauche: A French term used to denote a rough or blank timepiece *movement* without its *escapement*, *dial* or hands. (See page 115.)

Electric watch: A wristwatch which uses the electromagnetic principle and mechanical switches to impart *impulse* to a *balance* or motor and drive the watch time display.

Electronic watch: A wristwatch which uses electronic components (diodes, resistors, etc.) to provide an electronic switching action to control and drive a time display (either *digital* or *analogue*). The electronic switching is non-mechanical and therefore does not generate heat or frictional wear as the mechanical systems do.

Elliptical gear: A cam in the shape of an ellipse generally used to vary the length of a lever or arm.

Equation of time: The mathematical relationship between *solar time* and *mean time*. Usually found in chart form, table form or in an equation clock showing the difference between solar time and mean time on a particular date.

Equation timepiece: A clock or watch showing the difference between *solar time* and *mean time*, or showing the actual *solar time* and *mean time* at the same time.

Escapement: The system by which the *controller* of a timepiece allows the *gear train* to escape by equal, even amounts in order to accurately record the passage of time. Numerous escapement systems have been developed since the first *crown/verge wheel system* in the thirteenth century. Those that were successful over the years were the crown/verge, *anchor*, Graham dead-beat, pin wheel, Brocot, grasshopper and gravity escapements in clocks; and the *crown/verge*, *cylinder* and a series of detached *lever* escapements in watches: the rack lever, English lever, the anchor or Swiss lever, the *detent* and the *co-axial* escapements. (See Crown-Verge Escapement, page 22; Anchor Escapement, page 32; Swiss Detached Lever Escapement, page 81; Co-axial Escapement, page 79.)

Escape wheel: The last wheel in the *gear train* which is alternately locked and unlocked by the *controller* and at the same time gives *impulse* to the *controller* to keep it in motion, allowing the gear train to escape at a speed governed by the controller.

Établissage: A Swiss method of making watches whereby the parts are fabricated elsewhere, brought together and then assembled.

Établisseur: A workman who assembles watches for a *fabricant* in an *établissage*.

Fabricant: A *fabricant* in Switzerland is an assembler of watch parts on a large scale. The work includes finishing, dialing and casing the watch.

Fabrique: The place where a *fabricant* puts watches together. It roughly corresponds to a factory, but can be smaller than the word factory implies. In the eighteenth century the *Geneva Fabrique* was more a system (rather than a building) of manufacturing

EVOLUTION OF ELECTRONIC WATCH OSCILLATING FREQUENCIES

		Standard oscillating frequency
1952	Lip & Elgin electric watch prototype	5 beats (vibrations) per second
1957	Hamilton electric watch	5 beats (vibrations) per second
1960	Bulova Accutron	360 vibrations per second
1969	Seiko quartz analogue watch	32,768 vibrations per second

The accuracy of the watch improved as new oscillators were found that proved to be progressively more and more stable in their oscillating frequencies.

within the city from *ébauche* to finished watch, akin to the *cottage system* of manufacture

Finishing watchmaker: A highly skilled watchmaker who fits the wheels to a rough *movement* or *blanc*, installing *jewels* where needed; polishes the wheels, *arbors*, *pivots* and *plates*; fits the *escapement* and *balance* and adjusts the *pallets*, *hairspring* and *balance*, bringing all into smoothly running, consistent timekeeping.

Floor clock: A floor clock is any clock that stands on the floor. If it is in the hall, it is a hall clock. A floor clock is also known as a tall clock or a tall-case clock (in America) and a long-case clock (in the British Isles). A grandfather clock is a colloquial name in America for a striking and/or chiming floor clock that is over 6 feet (1.8m) tall. A grandmother clock is a shorter version of a grandfather clock, less than 6 feet (1.8m) tall.

Foliot: The first form of *controller* to be applied to a clock *escapement*. It has two arms with an adjustable weight at each end to bring the clock into time. The arms swing horizontally.

Form watch: A watch case in the shape of an animal, flower, cross, star or character.

Frequency: The number of swings, oscillations or vibrations that a *controller* makes in a specified period of time. *Balance wheels* and *pendulums* are usually stated as oscillations or "beats" per hour; *tuning forks* and *quartz crystals* and atoms are usually stated in vibrations per second.

Frequency oscillator: A *controller*. Can be any of the oscillators that provide a *frequency standard* for a timepiece.

Frequency standard: A constant, stable *oscillator* in a timepiece. A *controller* that provides the timekeeping performance standard or rate for the timepiece.

Fusée: A cone-shaped device that automatically compensates for the variable strength of a *mainspring*, keeping the effective torque on the *gear train* more or less equal. Used into the late nineteenth century when *mainspring* metallurgy improved to the point that *mainspring* power was much more consistent from a fully wound through to a nearly unwound condition.

Gear: A toothed wheel or cylinder which meshes with another toothed element to transmit power or change speed or direction. Also known as a *gear wheel*. In timepieces, a toothed wheel drives a toothed cylinder (called a *pinion*) that is connected directly to the adjacent wheel. Most mechanical timepieces utilize an intermittent flow of power through the *gears*, caused by the intermittent release of power by the *escapement* as the *gear train* drives the *oscillator*. Timepieces utilize a tooth shape called **epicycloidal** for wheel teeth and **hypocycloidal** for *pinion* teeth (called leaves). These related tooth forms or shapes provide for a smoother transmission of power through the gears of a timepiece than would the **involute** gear forms used in other machinery. **Involute gearing** is a stouter, stronger tooth shape that is much more suitable to the constant, uninterrupted power requirements of machines driven by constant, uninterrupted power (like a motor or steam engine). (See page 23.)

Gear form: The profile (or shape) of a gear tooth.

Gear train: A system of interconnected gear wheels. Also known as a gear wheel train and wheel train. (See page 23.)

Gear wheel: see *Gear train*.

Gilding: The art or process of applying a thin layer of gold to a metal surface. In the past, gold was applied to brass watch *movement plates* and *bridges* using gold powder mixed with mercury to form a paste that was brushed on, washed and then fired over charcoal to remove the mercury. The process is called fire gilding or water gilding and is very dangerous to execute. Many watchmakers were found slumped over their charcoal fires, dead from breathing the highly toxic mercury fumes. Electrogilding, or electroplating, is used today, except in restoration work which still requires fire gilding to restore the old gilding treatment. (See picture, opposite page.)

Gimbals: A device made of two rings mounted on axes at right angles to each other so that an object, such as a ship's compass or ship's *chronometer*, will remain suspended in a horizontal plane regardless of the ship's motion. (See page 137.)

Grande complication: A mechanical timepiece with numerous functions in addition to timekeeping, calendar, chronographic and *repeating* functions. See *Complication*.

Grande sonnerie: A form of quarter-hour striking in which the last hour struck is also repeated at each quarter, so the hearer can tell the time without seeing the *dial* of the timepiece.

Graver: A hand-held tool used to cut and shape metal on a revolving lathe. Also used by engravers to cut and chase designs on metal surfaces.

Hairspring: See *Balance spring*.

Hands: Mechanically driven index fingers on timepiece *dials* (or faces) that point to the time, lapsed time and other measurement indices.

Hand-wound watch: A mechanical watch *movement* whose *mainspring* is wound by hand, either by turning a *key* or a *crown* (or winding knob). The watch usually needs to be wound once each day, but there have been watches made that require winding twice a day, once every three or four days and even those that require winding once in eight days.

Horology: Has two distinct meanings: the science, study or theory of measuring time; and the art of making timepieces. From the Greek *hora* (hour) and *–logy* (subject of study). Related terms: **horologist** or **horologer** or **`orologer** or **`orologist**, meaning one who studies time measurement and/or its history; or one who makes timepieces. **Horologe** means a timepiece.

Impulse: The small force that is applied, via the *escapement*, to a *pendulum* or *balance wheel* to keep it swinging (or oscillating). Same as *impulse force*.

Impulse force: The force applied to the *oscillator* by the *escapement*, impulsing or transmitting energy to maintain the back and forth motion of the *oscillator* (*controller*).

Indexing system: *Escapement* system that counts the oscillations of the *controller* and releases the power of the *gear train*. The index, itself, is that part of the escapement that does the actual counting of the oscillations: *anchor/escape wheel* in a clock; *lever/escape wheel* or *detent/escape wheel* in a watch. Even the *quartz watch* has an indexing system, but it is done electronically within the circuitry of the quartz movement. (See Anchor Escapement, page 32.)

Isochronism: From *iso-* meaning *the same* and *chrono* meaning time. In a clock a pendulum's arc tends to be slowed by gravity, air resistance and an unwinding *mainspring*. In a watch the *balance wheel* has a long arc of swing when the watch is in the horizontal or dial position and a shorter arc in the vertical position. So *isochronism* is the ability of a *balance* or *pendulum* to traverse arcs of varying amplitude in the same length of time. This ability can be affected by the configuration and composition of the hairspring in watches, or the suspension spring in clocks. *Isochronism* is considered a precision adjustment for the accurate timepiece, along with positional adjustments, temperature adjustments and adjustments for barometric pressure.

Jewel: A natural or synthetic gemstone bored with a hole to act as a nearly friction-free bearing surface for a moving *wheel arbor pivot,* or as a perfectly flat surface to act as a *thrust bearing* on the end of a *wheel arbor pivot*. (See The Principles of the Watch, page 93.)

Key: see *Crown*.

Lapsed time recorder: A feature of some *chronographs* and stopwatches that records lapsed time in minutes and hours using sub-dials and indicator hands. Some have minute-recorder sub dials and some have both minute- and hour-recorder sub dials.

Lathe: A machine on which a piece of metal is spun and shaped using hand-held *gravers* or fixed cutting tools and abrading tools.

Lépine movement: A style of watch *movement* that originally had a separate *bridge* for each wheel in the movement. Now it is a Swiss term for a thin, open-faced watch; i.e., a watch without a metal cover over its face. It can be used for either a pocket watch or a wrist watch.

Lever escapement: The most successful of all *escapements*, invented by Thomas Mudge in 1759 and improved by Josiah Emery (later part of eighteenth century into early nineteenth century) in 1785. It is also called a "detached lever" because it detaches itself from the *balance wheel* during much of the balance wheel's oscillation. The lever escapement uses a Y-shaped lever (called a "*pallet fork*") that rocks back and forth to the rhythm of the balance wheel's oscillation, locking and releasing each tooth of the escape wheel, counting the oscillations of the balance wheel, releasing the wheel train of the watch and at the same time maintaining the oscillations of the balance wheel by imparting an *impulse* to the balance wheel with each oscillation.

Lignum-vitae: An American tropical tree (*Guaiacum officinale or G. sanctum*), either of which species has a very oily and dense wood grain structure. The wood is self-lubricating and so is used for pivot *bearing holes* and other components in wooden-works clocks and some notable precision timepieces.

Locking plate: A "counting wheel" for a striking clock. Notches in the rim of the locking plate or *count wheel* determine the number of hammer blows the *striking mechanism* of the clock will deliver to the gongs or bell of the clock.

A gilt- or gold-plated, brass miniature tower table clock with automata and a carillon of tuned bells, made by Lucas Weydmann, Krakow, 1648.

Lunar watch: A type of *astronomical timepiece* that can show the phases of the moon, the age of the moon and/or its journey through the zodiac.
Luxury watch: A category of watch marketed as much for its prestigious name as for its fine quality. Designed as a symbol of extravagant living and as a status symbol, luxury watches used to be encrusted with diamonds and other gemstones, but today denote high quality watch *movements* enclosed in highly finished, finely crafted *cases* and bands of precious metals and combinations of precious metals and stainless steel. Luxury watches are generally marketed through upscale jewelry and department stores.
Mainspring: A coiled ribbon of steel tightly coiled around the main *wheel arbor* of a timepiece which provides the motive force for the *gear train*. Early *mainsprings* released their power unevenly: strong power when fully wound and weaker power as the *mainspring* unwound. This inconsistency brought about power compensation devices to even out the release of power, such as the *stackfreed* and the *fusée*. Also used was the stopworks, a mechanism designed to use just the middle range of power by preventing the spring from being fully wound or fully unwound. (See Stackfreed and Fusée box, page 36.)
Mainspring barrel: A cylindrical container that often holds a *mainspring*. The first wheel of the *gear train* is often a part of the container, but not always.
Mean solar day: The average *solar day*, divided into 24 equal hour periods with each hour divided into 60 equal minutes and each minute divided into 60 equal seconds. Also known as clock time and *mean time*.
Mean sun: A hypothetical sun defined as moving at a uniform rate along the celestial equator so that it completes its orbit in the same period as the apparent sun, used in computing the *mean solar day*.
Mean time: Time measured with reference to the *mean sun*, giving an equal 24 hours throughout the year.
Meridian: A great circle on the earth's surface passing through both poles, represented on a globe by lines running from the North Pole to the South Pole. When the sun shines directly down on a meridian line, it is noon all along that meridian. There are an infinite number of meridian lines. Wherever you stand on the earth, a meridian line passes through that spot. But scientists have arbitrarily divided the earth into 24 principal meridians each one 15 degrees from the next one (or one hour of mean solar time distant) and measured from the Prime Meridian that runs through Greenwich, England, so that we may locate ourselves and our activities at any one specific point at any one specific time anywhere in the world.
Module: The framework of the electronic timepiece with its electronic components and *frequency oscillator*. Also known as the *works* of the timepiece.
Movement: The framework of the mechanical timepiece, complete with its *wheel train, mainspring, escapement* and *frequency oscillator*. Also known as the *works* of the timepiece. The module of an electronic timepiece is often mistakenly called a movement.
Oil sink: The cupped depression around a *pivot bearing hole* that retains a supply of lubricating oil for the *arbor*.
Old timepiece: A timepiece made prior to the present generation (of humans, not timepieces). Includes all ages of timepieces: *ancient, antique* and vintage.
Oscillator: The *frequency* standard of a timepiece. Also known as the *controller*. Any device that vibrates at a fixed frequency, *pendulum, foliot, balance, tuning fork, quartz crystal,* molecule or atom. (See Crown-Verge Escapement, page 22.)
Pallet fork: The "Y"-shaped lever of the *lever escapement* that *indexes* the power of the *gear train* of a timepiece, both *impulsing* the *oscillator* and counting its oscillations. *Pallets,* or angled impulse surfaces on the pallet fork, slide along the angled teeth of the *escape wheel* to transmit the impulse to the *oscillator*. (See Swiss Detached Lever Escapement, page 81.)
Pendulum: A shaft pivoted or suspended from a cord or spring that oscillates back and forth at a fixed rate. (See Anchor Escapement, page 32.)
Perpetual calendar: A *complication*, this calendar mechanism automatically adjusts for 30-day months and leap years.
Petite sonnerie: A timepiece that strikes just the hour.
Piercing saw: A jeweler's saw using a very fine saw blade to pierce or cut intricate shapes and patterns of holes into flat metal objects.
Pin set: A mechanism to disengage the *crown* or winding knob from its *mainspring* winding function; and simultaneously engage the hands so that they may be reset to the correct time. An evolution between the key setting mechanism and the stem setting mechanism, it was used from around 1870 to 1910. (See caption, page 113.)
Pinion: A toothed cylinder used to connect adjacent toothed gears in a *gear train*. The teeth of the pinions are called **leaves.** (See How a Watch Works, page 93.)
Pivot: The small end of a shaft or *arbor* that runs in a *bearing hole*. The smaller diameter of the pivot helps reduce running friction. The shoulder between the larger diameter of the arbor and the smaller diameter

of the pivot keeps the arbor in place between the *plates* and *bridges* of the timepiece. (See The Principles of the Watch, page 93.)

Pivot bearing holes: The hole in which the pivot bearing turns.

Plates: The front and back surfaces of a timepiece *movement*, joined by posts or **pillars,** making up the timepiece's frame. Within these plates are the *pivot bearing holes* in which the *gear wheel* shafts turn. In watches the main plate (front plate) may have *bridges* instead of a back or upper plate to make up the framework. The main or front plate has the *dial* affixed to it; the back plate (or bridges in later movements) is commonly decorated with *damascening*, engraving or other decorative treatments. (See picture below.)

Positional timing: A watch runs at different *rates* in different positions because of errors inherent in the *escapement* and changes in the center of gravity of the *balance* and/or its *hairspring*; and errors in *isochronism* as the back-and-forth arcs of the balance shorten as the *mainspring* runs down and the balance position changes. The timing *rate* while the watch is worn is an average of these errors and therefore varies with different wearers. A good wrist watch is tested and adjusted in six positions: with its dial up; dial down (the horizontal positions); and crown up, crown down, crown to the left and crown to the right (the vertical positions). Positional errors affect a pocket watch, as well, but usually it needs only to be tested and adjusted in four of its positions: it moves in the pocket in the vertical positions of crown up, crown left and crown right, but seldom in the crown down position. The pocket watch runs at night in the dial up position, but seldom in the dial down position. Positional error is not a factor in clocks since they always remain in an upright position.

The front and back plates hold the wheels, arbors and levers of this Greenwich Royal Observatory regulator clock; made by Thomas Tompion, London, 1676.

Precision timekeeping: A clock or watch designed and adjusted to keep accurate and consistent timing rates despite temperature and barometric changes, positional influences on the *oscillator* and *isochronal* errors and errors inherent in the *escapement*, gear time and power source.

Quartz crystal: A naturally occurring mineral which, when sliced thinly and energized with an electric current, will oscillate, or vibrate, at a very high, constant *frequency*.

Quartz watch: An electronic watch that uses a *quartz crystal* as its *frequency standard*. It can use either a *digital* or *analogue* format to display the time.

Rate: The timekeeping performance of a watch or clock as measured against a *time standard*. Measured in seconds per day and, in the case of a watch, measured and adjusted in the various positions in which the watch may run. A gaining rate means the timekeeper is running faster than the time standard that it is measured against.

Rating certificate: An official certificate attesting to the timing performance of a timepiece. There are two main classes of rating certificates. One is for special timekeepers and is issued by the National Physical Laboratory, England; the Geneva Observatory, Switzerland; and the German Hydrographic Institute. The other is for Swiss production watches of high performance. These certificates are issued by one of the official bureaus for testing watches in Switzerland. These certificates accompany the watches when they are sold. By Swiss law, only watches gaining certification can be called "Chronometers."

Recoil: Another name for an *anchor escapement* because the *gear train* wheels turn backward slightly after every jump forward. *Recoil* can be easily seen by watching the second hand of a long-case clock.

Regulator: (a) A very accurate long-case clock, adjusted for temperature and barometric pressure, and without a *striking* or chiming mechanism or other *complication* that might interfere with the accuracy of the clock. Also known as a **jeweler's regulator, watchmaker's regulator,** or a **scientific regulator.** (b) A mechanical device for adjusting the timing *rate* in a timepiece. (c) A **patented regulator** is a device

EVOLUTION OF THE QUARTZ WATCH

	Date	Power source	Oscillator	Indexing
Mechanical	(1600s to present)	Mainspring	Balance wheel	Pallet fork releasing an escape wheel
Electric	(1957 –1960s)	Battery	Balance wheel	Balance/motor with electromechanical contacts driving an index wheel
Electronic	(1970s)	Battery	Balance wheel	Electronic circuit indexing a motor
Accutron	(1960 –1970s)	Battery	Tuning fork	Oscillator driving an index wheel
Quartz analogue	(1969 to present)	Battery	Quartz crystal	Electronic circuit indexing a motor

Of the five major components of the mechanical watch, only the hand-setting mechanism and the dial and hands remained the same when transferred to most electrical watches. The internal components changed: the power source, the oscillator (frequency standard), and the indexing system (i.e., the way the frequency is counted or divided into one-second intervals). In the digital or solid-state quartz watch, even the hands, dial and setting mechanism changed.

for adjusting the timing *rate* in a watch and for which design patents were granted to numerous inventors around the world from the late seventeenth century to the present day.

Repasseur: A Swiss term for a *finishing watchmaker* who finishes and adjusts the *blanc* or *blanc roulant.*

Repeating mechanism/Repeater: A form of *complication* that enables a clock or watch to strike the last hour and (depending upon its type) the nearest quarter hour and even the nearest minute upon demand by pulling a cord or pushing a lever.

Restoration: In addition to repair, restoration involves techniques and measures to return the appearance and function of the timepiece to its original (or perceived original) condition. In doing restoration work the watchmaker or clockmaker often has to resort to machinery, tools and techniques used when the timepiece was originally made.

Self-winding watch: An *automatic* mechanical watch.

Sidereal time: The time elapsing between two successive transits of a fixed star across a *meridian.* A sidereal day is 3 minutes, 56 seconds less than a *mean solar day.*

Size: An American term for the *calibre* of a watch movement. Uses the Old Lancashire, England, system of movement measurement. In this system an 18-size pocket watch movement (written 18/s) is 1.766 inches in diameter or 44.86 millimeters in diameter.
Other examples:
16/s = 1.700 inches or 43.18 millimeters
6/s = 1.366 inches or 34.70 millimeters
0/s = 1.166 inches or 29.62 millimeters

Skeletonized: A watch *movement* whose *plates* and *bridges* have been pierced to expose the *gear train* wheels, *escapement* and *oscillator.* Usually enclosed in a case with a see-through *dial* and glass back.

Skew gearing: Having a part or *gear* that diverges from a linear path.

Solar day: The time it takes the sun to return to the same position from one day to the next. The elapsed time for this observation is called a real solar day. The length of a real solar day varies from season to season.

Solar time: Sun time. The time measured by a sun dial. A *solar day* is from the sun's highest point in the sky on one day to its highest point in the sky the next day. Solar days vary throughout the year. A clock measures average or *mean solar time* and divides it into equal hours.

Solid state: An *electronic watch* that has no moving parts and a *digital display* of time.

Spiral spring: See *Hairspring*.
Split-second timer: A timer with two second hands, one superimposed over the other and independently controlled to record two events at one time.
Stackfreed: A cam and spring arrangement that attempts to equalize the varying torque of the old hand-made *mainsprings*. Used primarily in *ancient* German portable timepieces. (see Stackfreed and Fusée box, page 36.)
Stopwatch: A timer for recording lapsed time, but without a timekeeping mechanism to measure the time of day. Also known as a sports timer or production timer.
Striking mechanism: A mechanism that sounds the hours on a bell or gong, mechanically remembering the current hour's strike on a *locking plate* or *count wheel*.
Striking train: A series of *gear wheels* that drive a hammer that strikes a bell or gong. The striking train is released on each hour by the time train, the duration of which is determined by the *locking plate* or *count wheel* that remembers the next sequence of hammer strikes needed and locks the striking train when the correct strike is finished.
Temperature compensation: A method of avoiding or obviating timekeeping variation (temperature error) at different temperatures, which affects *pendulums, balance wheels, hairsprings, quartz crystals* and most other *time standards*. Compensation pendulums are designed to keep the pendulum lengths from changing. Balance wheels and hairsprings are each affected separately and differently, by temperature changes. The pendulum, balance wheel and hairspring are each compensated by integrating various alloys and methods into their design and construction.
Temporal hours: Or temporal time. In ancient times daylight and nighttime were split into a number of hours, usually 12 each. A daylight hour was different than a nighttime hour and both varied at different times of the year, but not by much in the Mediterranean where most early civilizations thrived.
Thrust bearing: A bearing plate upon which the end of an *arbor* or axle rides. The thrust bearing limits the lateral movement of an arbor or axle.
Time: A continuum of events that occur in apparently irreversible succession from the past through the present and into the future. The passage of time is measured by referring to recurring phenomena, such as the rotation of the Earth, *oscillating pendulum, balance wheel, quartz crystal,* molecule or atom. Two kinds of time are usually used: the time of day, measured by the rotation of the Earth; and the duration of an interval of time, measured by an accurate timepiece.
Time standard: A constant, stable timepiece against which other timepieces are measured. At one time it was a fixed star and the sun; then it was a precision mechanical *regulator*; now it is a quartz clock and an atomic clock.
Tourbillon: A special watch *escapement* mounted on a platform that revolves, usually once a minute, evening out some of the positional errors inherent in

THE LINEAGE OF PRECISION-REGULATOR CLOCKS

	Error per day	
1500	Foliot oscillator	15 seconds
1650	Huygen's pendulum clock	10 seconds
1700	Improved escapements	8 seconds
1721	Graham's mercury pendulum	1 second
1726	Harrison's gridiron pendulum and reduced friction	0.33 second
1921	Shortt's free pendulum	0.004 second
1948	Quartz crystal clock	0.0003 second
1955	Cesium atomic clock	0.000005 second

The accuracy of the clock improved as new oscillators were found that were progressively more and more stable in their oscillating frequencies.

the ordinary watch *escapement*. (See diagram below.)
Train: a series of interconnected gear wheels, also known as a *gear wheel train* and a *wheel train*. (See The Principles of the Watch, page 93.)
Trundle: Another name for a *pinion* gear; but usually with rollers instead of gear teeth (leaves).
Tuning fork: An electronic watch *frequency standard* that *indexes* and drives a *gear train*. Developed and introduced by the Bulova Watch Company and later licensed to a Swiss *ébauche* manufacturer, the tuning fork electronic watch was originally marketed under the Accutron name. (See How the Electrical Watch Works, page 155.)
Turns: An elementary form of *lathe* used for centuries by watchmakers and clockmakers and even today for restoration work and repair of watches. It consists of a frame with two fixed points, or **centers,** between which the work is fixed. The work is then turned by a bow with a string wrapped around the work. In early times the work was turned with one hand and shaped with a *graver* in the other hand. But in later times the work was usually shaped in a foot-powered or engine-powered *lathe* and then burnished or polished in the turns to ensure the greatest precision.
Verge escapement: The earliest *escapement,* first known in the thirteenth century and made until the late nineteenth century. The *balance staff*, or verge, has two *pallets*, about 95 to 100 degrees to each other, which alternately escape the teeth of a crown-shaped *escape wheel*, called a *crown wheel*. (See Crown-Verge Escapement, page 22.)
Watch: A portable clock designed as a personal timepiece that could be worn in the pocket (a pocket watch), around the neck (a pendant watch), on a finger (a ring watch) or on the wrist (a wrist watch).
Watchmaker: One whose occupation is making or repairing watches. A watch repair technician is one who can replace defective component parts with previously purchased replacement parts and adjust the performance of the watch.
Wheel arbor: A horological name for a wheel shaft or axle in a clock or watch; also see *Arbor*. (See The Principles of the Watch, page 93.)
Wheel arbor pivot: The end of a *wheel arbor*, smaller in diameter than the *arbor* that runs in a *bearing hole*; also see *Pivot*. (See The Principles of the Watch, page 93.)
Wheel pivot bearing holes: A hole in which a *wheel arbor* rotates; see *Bearing hole*. (See The Principles of the Watch, page 93.)
Wheel train: see *Gear train*.
Winding stem: A shaft that connects the winding *crown* or knob with the winding mechanism of the watch. Evolved from the **winding square** that needed a key to turn it and its winding mechanism. *Winding stems* then evolved into a multi-functional shaft that linked the *crown* to both the winding mechanism and the hand-setting mechanism and even the latch to unlock the front cover on a covered pocket watch. (See The Principles of the Watch, page 93.)
Works: See *Movement*.
Zodiac: The band of sky through which the sun, moon and the principal planets appear to move around the Earth. It is divided into 12 segments of 30 degrees each. Each segment is known by its own constellation of stars, which represents a mythological character and is known as one of the 12 signs of the zodiac.

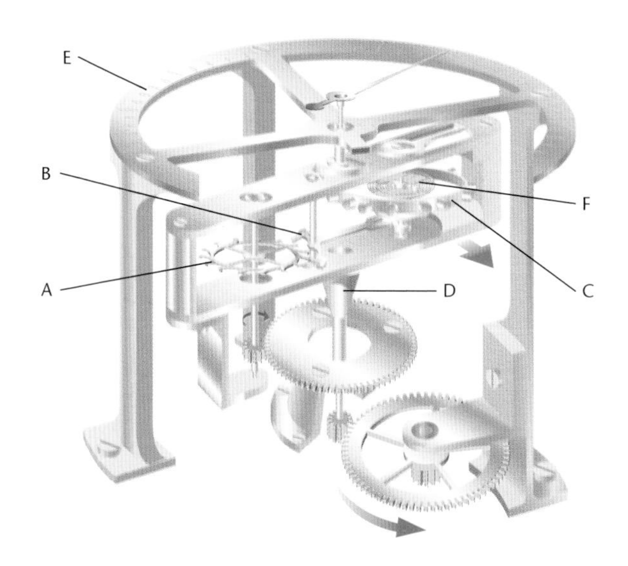

LEFT
The tourbillon simply puts the escapement (escape wheel A with its pallet fork B) and the balance wheel (C) into a cage (E, called a tourbillon) that revolves on the fourth wheel (D), i.e. the wheel that turns the second hand one revolution in a minute. This means that the balance wheel (C) is carried in a 360-degree circle each minute, eliminating variations in timekeeping rate that are caused by gravity's pull on the balance wheel, balance spring (F) and pallet fork (B). The concept is simple but it is the most difficult escapement system to make and get right.

SOURCES AND FURTHER READING

Timepieces: Masterpieces of Chronometry is more than a history of timekeeping: it is a story of timepiece artistry, beauty and technology, set into the context of the religious, commercial and social fabric of our lives. This account is not a scholarly study, so I have not used footnotes or annotations ... but it is factual and accurate.

A number of books and articles have influenced my view of timekeeping history and I have listed those sources that I have drawn upon to tell this story. If this story has piqued your interest for more information, I'd like to refer you to these various sources for your further reading:

Andrews, William J.H., ed. *The Quest for Longitude.* Cambridge, Mass.: Collection of Historical Scientific Instruments, Harvard University, 1996.

______ **and Seth Atwood.** *The Time Museum: An Introduction.* Rockford, Illinois: The Time Museum, 1983.

Babel, Anthony. *Histoire Cooperative de l'Horlogerie, de l'Orfevrerie* (appendix II). Geneva: A. Julien, 1916.

Bailey, Chris. *Two Hundred Years of American Clocks & Watches.* Englewood Cliffs, New Jersey: Prentice-Hall, Inc., 1975.

Baillie, G. H., C. Clutton and **C. A. Ilbert.** *Britten's Old Clocks and Watches and their Makers*; 7th edition. New York: Bonanza Books, 1956.

Cardinal, Catherine. *Watchmaking in History, Art and Science: Masterpieces in the Musée International de'Horlogerie, La Chaux-de-Fonds, Switzerland.* La Conversion: Editions Scriptar S.A., 1984.

Christianson, David A. "Commercial Evolution of the Watch." Lecture presented to the New Orleans National Convention of the National Association of Watch and Clock Collectors, July 13, 2001.

Clutton, Cecil, and **George Daniels.** *Watches.* New York: The Viking Press, Inc., 1965.

Fried, Henry B. "Invention's Debt to Horology." A lecture first given before the New York Chapter of the National Association of Watch and Clock Collectors and then prepared as a slide presentation for the American Watchmakers Institute, @ 1975.

______. *The Museum of the American Watchmakers Institute.* Cincinnati: AWI Press, 1993.

Harrold, Michael C. *American Watchmaking: A Technical History of the American Watch Industry 1850–1930.* Columbia, Pennsylvania: NAWCC, 1984.

Jaquet, Eugene, and **Alfred Chapuis.** *Technique and History of the Swiss Watch.* London: Spring Books, 1970.

Landes, David S. *Revolution in Time.* Cambridge, Massachusetts: Harvard University Press, 1983.

Leuthner, Stuart. "Star Calibre 2000," in *Watch & Clock Review,* November/December 2000, pp. 42-46.

______. "The War of the Complications," in *Watch & Clock Review,* April 2000.

Matz, Benjamin. *The History and Development of the Quartz Watch.* Denver: *Chronos* Magazine, 1999.

McKinnie, M.P. "Webb C. Ball Watches," in *Bulletin of the National Association of Watch and Clock Collectors,* Volume 14, Number 2; Whole Number 144, February 1970, pp. 180-183.

Nicolet, Jean-Claude. "The Search for the Perfect Escapement," in *American Time,* December 1999/January 2000, pp. 54-58.

Pratt, Derek. "George Daniels – Man of Action," in *Horological Journal,* August 1996, pp. 254-255.

Price, Derek J. de Solla. "Clockwork Before the Clock and Timekeeper Before Timekeeping," in *Bulletin of the National Association of Watch and Clock Collectors,* Volume 10, Number 12; Whole Number 106, October 1963, pp. 950-961.

Roberts, Kenneth D. *The Contribution of Joseph Ives,* Revised Second Edition. Fitzwilliam, New Hampshire: Ken Roberts Publishing Co., 1988.

Richardson, John M. "Time and Its Inverse," in *Bulletin of the National Association of Watch and Clock Collectors,* Volume 10, Number 8; Whole Number 102, February 1963, pp. 597-604.

Robinson, Mary Frances, Ph.D. "The Geneva Rules of 1601," in *The WatchWord,* a publication of the North Carolina Watchmakers' Association, March/April 2001.

Roehrich, Jean Louis. "The Cabinotier," in *National Jeweler,* 1958.

______. "The Vallée de Joux: Cradle of Complicated Watch Mechanisms," in *National Jeweler,* September 1959.

Stephens, Carlene E. *Inventing Standard Time.* Washington, D.C.: Smithsonian Institution, 1983.

Stephenson, Bruce, Marvin Bolt and **Anna Felicity Friedman.** *The Universe Unveiled: Instruments and Images Throughout History.* Chicago: Adler Planetarium & Astronomy Museum; and New York: Cambridge University Press, 2000.

Tardy. *Dictionnaire des Horlogers Français.* Paris: Editions Tardy, 1972.

Thompson, Joe. "Patek Philippe's 'Calibre '89': The Making of a Master Timepiece," in *Modern Jeweler,* April 1989.

Ward, F.A.B. "The Evolution of Engineering: How Timekeeping Mechanisms Became Accurate," in *Bulletin of the National Association of Watch and Collectors,* Volume 10, Number 12; Whole Number 106, October 1963, pp. 950-961.

Whitney, Marvin E. *The Ship's Chronometer.* Cincinnati: AWI Press, 1985.

INDEX

COMMERCIAL EVOLUTION OF THE WATCH

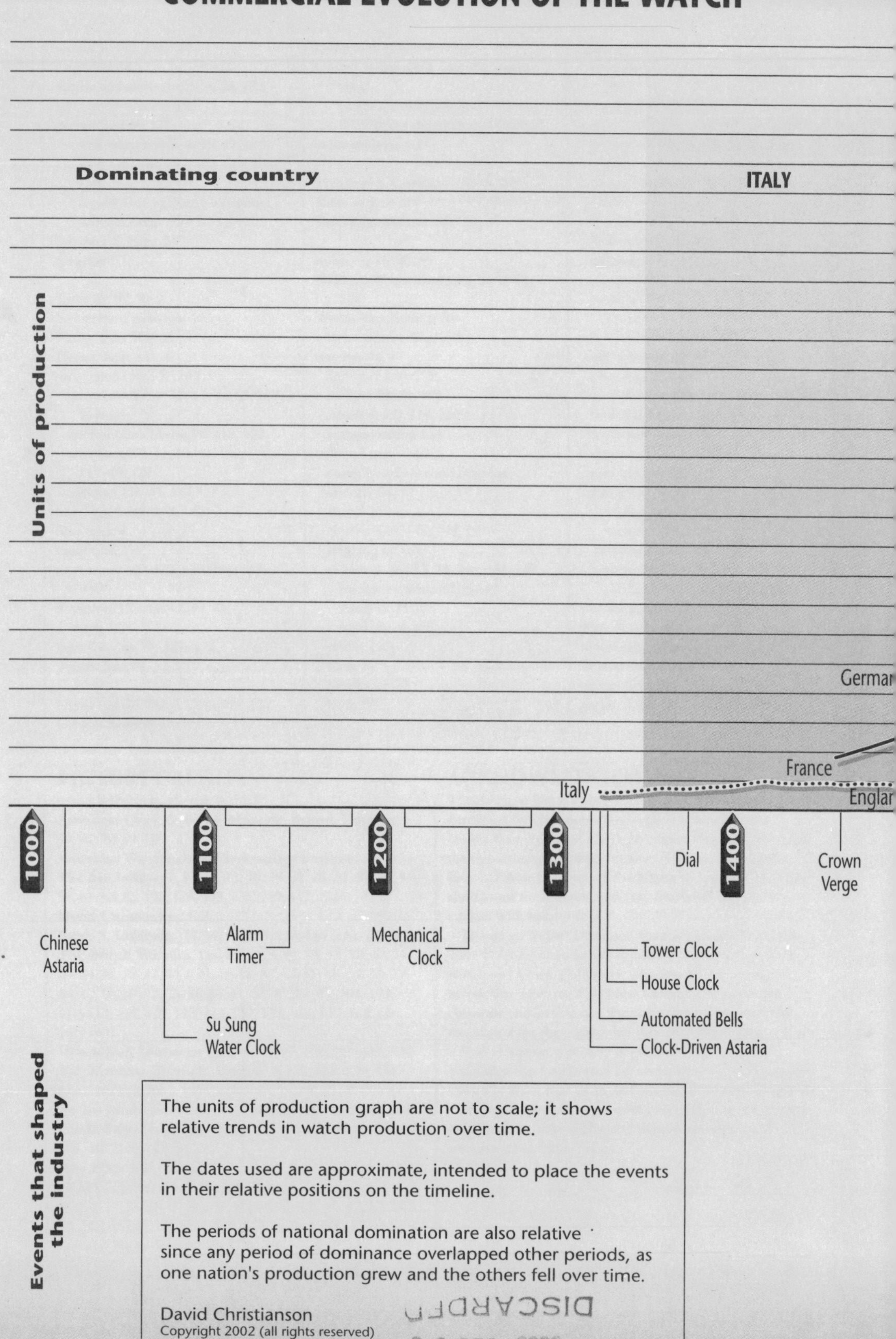

The units of production graph are not to scale; it shows relative trends in watch production over time.

The dates used are approximate, intended to place the events in their relative positions on the timeline.

The periods of national domination are also relative since any period of dominance overlapped other periods, as one nation's production grew and the others fell over time.

David Christianson